The Mathematics of Elections and Voting

T0155759

W.D. Wallis

The Mathematics of Elections and Voting

 Springer

W.D. Wallis
Department of Mathematics
Southern Illinois University
Evansville, IN, USA

Additional material to this book can be downloaded from http://extras.springer.com

ISBN 978-3-319-09809-8   ISBN 978-3-319-09810-4 (eBook)
DOI 10.1007/978-3-319-09810-4
Springer Cham Heidelberg New York Dordrecht London

Library of Congress Control Number: 2014950633

Printed on acid-free paper

Springer is part of Springer Science+Business Media (www.springer.com)

for Madelin Marie

Preface

In recent years, there has been an increase in the number of college courses designed to introduce mathematics to non-math students. One of the topics usually included is a very cursory introduction to the theory of voting. This is a topic that involves some interesting mathematics, although there have not been very many advanced papers on the subject.

I thought I would like to introduce voting to more advanced students of mathematics. This volume is designed for such students. I have introduced a number of electoral methods and included a number of exercises. However, I assume those who are really interested in this area of mathematics will explore the topics further online and in other sources, such as [28] and [31].

There are several worked examples in the text, and some are followed by practice exercises. I would suggest that readers go through these as they encounter them. The exercises are at the end of the chapters, and answers to the odd-numbered questions will be found at the end of the book. I have supplied more complete solutions to many of them than is usual in a text. But please try them before you look at the answers. A complete solutions manual is also available.

I would like to express my thanks to the staff of Springer, and in particular to Razia Amzad for her help and encouragement.

Evansville, IN, USA   W.D. Wallis

Contents

Chapter 1

Introduction

There are many situations that call for a group decision. At one extreme, three friends might be trying to decide where to go for dinner. At the other end of things, millions of people often need to decide which individual, or which political party, will lead their country. Very often we decide by *voting*. But what is the best—fairest, most representative—voting system? This is more complicated, and less obvious, than you might think.

When there are just two candidates for a post, it is all very simple: just vote for the person you prefer. But, as soon as there are more than two candidates, confusion arises. Remember the 2000 Presidential election? Many people believe the presence of Ralph Nader on the 2000 ballot affected the results.

1.1 Summary of the Text

In Chap. 2 we shall look at *simple* ballots, in which only one person is to be elected. We shall discuss majority and plurality voting, runoff elections and the Hare (or *instant runoff*) method and the Coombs method. We shall also look at elections where points are allocated to the candidates and the high scorer wins; sporting contests, where points are awarded not by electors (the judges) but based on the players' performance, are very similar to elections of that kind. We conclude by examining the way in which the United States President is elected.

We shall look at simple elections further in Chap. 3. In particular, we shall examine a technique proposed by the eighteenth century scholar Nicholas de Condorcet, to find a result when more than one possibility arises. We shall also study the *Condorcet winner criterion* and the *Condorcet loser criterion*, measures of the fairness of electoral systems that are based on the work of Condorcet. We shall also examine the Bucklin method and sequential pairwise voting.

© Springer International Publishing Switzerland 2014
W.D. Wallis, *The Mathematics of Elections and Voting*,
DOI 10.1007/978-3-319-09810-4_1

Not every simple election will have a result; it is always possible that the voters like two candidates equally well, and there is a tie. But we shall also see some methods that sound very reasonable, but do not result in the election of a candidate even when there are no ties. Moreover, it is possible that two different systems, each of which sounds perfectly fair, can give different results.

Sometimes groups of voters can manipulate the system to favor their candidates. This is particularly true when amendments to motions are being considered; in Chap. 4 we look at fair and unfair voting and manipulation techniques, with special reference to amendments. We also discuss polls, and the way in which polls do not only predict the outcome of elections, but also can influence the result.

You may ask what is the best, fairest method of election. In 1950, Kenneth Arrow put forward three very reasonable conditions of electoral fairness. He then proved that no voting system for more than two candidates can satisfy these conditions. We shall examine this result, called *Arrow's Impossibility Theorem*, and the later, more powerful *Gibbard-Satterthwaite Theorem*, in Chap. 5. One can also have a *complex* election, where more than one candidate is to be elected. For example, if a club is to have a president, a secretary and three committee-members, there may be two simple elections—one for president and one for secretary—plus a complex election for the other three members. Complex elections, which are also called *multiple elections*, will be discussed in Chap. 6.

While many systems either require electors to select just one candidate or else to rank the candidates in strict order. But there are some methods that allow ties; an example is *approval voting*, where voters simply divide the candidates into two groups, those who are considered suitable for election and those who are not. More complicated methods, such as *range voting*, are also used. We shall examine these more general systems in Chap. 7.

Although ties are unlikely in most political elections, because of the large number of voters, they are more common when a committee votes. To get around this problem, it is common for a committee chair to have a *casting vote*—effectively, the chair has a second vote when there is a tie. In 1954 Shapley and Shubik proposed a measure of deciding how much influence the chair had compared to the other committee members, called the *Shapley-Shubik Power Index*. There are more general cases where individuals have different numbers of votes. These can be analysed in the same way. A slightly more general case is that of *coalitions*. Several people agree to vote as one, but it is always possible that one or more will change their mind. The analysis of more general cases has led to another measure of power, the *Banzhaf power index*. Chapter 8 is devoted to these topics and to these two power indices.

In all, there is considerably more mathematical content in the study of voting and electoral systems than one might at first expect. We shall analyse the situation, particularly in Chaps. 5 and 8.

Chapter 2
Simple Elections I

The simplest form of election is where there are a number of candidates for office, and one is to be elected. There is a well-defined set of voters—the *electorate*—and each voter casts one vote. The votes are then counted. We shall refer to such an election as a *simple election*.

In this chapter we shall look at some of the systems that have been suggested for simple ballots and are used in various places.

2.1 Elections with Two Candidates

Suppose two people are running for an office. After each elector makes his or her one vote, the person who gets more than half the votes wins. This is called the *majority* or *absolute majority* method. If there is an even number of electors then ties are possible—ties are possible in any electoral system—but apart from this the absolute majority method always produces a result.

2.2 Majority and Plurality

If there are three or more candidates, the majority method is not so good; there may quite easily be no winner. Several schemes have been devised that allow a candidate with an absolute majority to be elected, and try to find a good approximation when there is no "absolute" winner. These are called *majoritarian* or *plurality* systems.

© Springer International Publishing Switzerland 2014
W.D. Wallis, *The Mathematics of Elections and Voting*,
DOI 10.1007/978-3-319-09810-4_2

The first generalization is the *plurality* or *simple majority* method; it is often called "first-past-the-post" voting. Each voter makes one vote, and the person who receives the most votes wins. For example, if there were 3 candidates, A, B and C, and 70 voters, the absolute majority method requires 36 votes for a winner. If A received 30 votes and B and C each got 20, there would be no winner under that method. Under plurality, A would be elected.

The problem with the plurality method is that the winner might be very unpopular with a majority of voters. In our example, suppose all the supporters of B and C thought that both these candidates were better-qualified than A. Then the plurality method results in the election of the candidate that the majority thought was the worst choice. This problem is magnified if there are more candidates; even if there are only four or five candidates, people often think the plurality method elects the wrong person.

To overcome this difficulty in countries with only two major political parties, it is common for each party to endorse only one candidate. For example, in the United States, if there are two or more members of the Republican party who wish to run for some office, a preliminary election, called a *primary election*, is held, and party members vote on the proposed candidates; the one that receives the most votes is nominated by the party, and usually the others do not stand for election. This election would be called the *Republican primary*. There usually will also be a Democratic primary, and sometimes other parties run primaries. However, this method will not solve the problems if there are several major parties, or if the post for which the election is held is not a political one.

For further discussion of simple majority elections, see the paper of May [18].

2.3 Sequential Voting

Another technique used to avoid the problems of the plurality method is *sequential voting*. In this scheme a vote is taken, as a consequence a new set of candidates is selected; then a new vote is taken. The aim is to reduce the set of candidates to a manageable size—often to size two. The original election is again called a *primary*, but in this case all the candidates run in the primary election, not just those in one party.

For example, when the city of Carbondale, Illinois, votes for Mayor, there is a primary election for all the Mayoral candidates. This is run like an ordinary election. Later there is another election; the candidates in the final election are the two candidates who received the most votes in the primary. This second (*runoff*) election is decided by the majority method. A similar method is used in many other cities in the United States and in electing the Presidents of France and Chile and a few other countries.

We shall refer to this as the *runoff method* or *plurality runoff method*. The two top candidates are decided by plurality vote; all other candidates are eliminated; then a majority vote is taken. In the very unlikely case of a tie, some modification is necessary. For example if two candidates are tied for second place, those two and the vote leader would be considered in the runoff after the votes for any eliminated candidates have been distributed. In some cases the candidate with the smallest number of votes would be eliminated—essentially, it would be treated as a second runoff—while in others the race between those three candidates would be decided by plurality.

In electing state Governors in the United States and Presidents of a number of other countries (particularly in Europe), a modified runoff method is used. Separate primary elections are conducted by the different political parties, and each party nominates only one candidate in the final election. A more complicated system is used in electing the American President; for details, see Sect. 2.7.

In the real world there is usually a delay after the primary, and more campaigning takes place. As a result of this, individual voters' preferences may change. But for simplicity's sake we shall ignore this for the moment, and assume that every voter has a *preference list*, an order of preference between the candidates that remains fixed throughout the voting process; we assume that all candidates are included and there are no ties. We define the *preference profile* of an election to be the set of all the voters' preference lists. This can conveniently be written in a table. For example, say there are three candidates, A, B, C, and suppose:

> 5 voters like A best, then B, then C;
> 7 voters like B best, then A, then C;
> 4 voters like A best, then C, then B;
> 3 voters like C best, then B, then A;
> no voters like B, then C, then A;
> no voters like C, then B, then A.

We can represent this as

5	7	4	3
A	B	A	C
B	A	C	B
C	C	B	A

(We could also write

5	7	4	3	0	0
A	B	A	C	B	C
B	A	C	B	C	A
C	C	B	A	A	B

but we shall usually omit zero columns.)

In the primary election, each voter votes for the candidate he or she likes best. In the example, A would receive nine votes—five from those with preference list ABC, and four from those with list ACB. In general, a candidate receives the votes of those who put that candidate first in the preference list.

Sample Problem 2.1 *Suppose the preference profile of an election is*

5	7	4	3
A	B	A	C
B	A	C	B
C	C	B	A

What is the result of the election in the following cases?

  (i) *The majority method is used.*

 (ii) *The plurality method is used.*

(iii) *The runoff method is used.*

Solution. A receives nine votes, B receives seven, C receives three. So (i) there is no majority winner (as there are 19 voters, 10 votes would be needed), and (ii) A is the plurality winner. In a primary election, A and B are selected to contest the runoff. For the runoff, C is deleted, so the preference profile is

5	7	4	3
A	B	A	B
B	A	B	A

,

or (combining columns with the same preference list)

9	10
A	B
B	A

So B wins the runoff.

Practice Exercise. Repeat this question for an election with preference profile

7	5	8	3	4
A	A	B	B	C
B	C	A	C	B
C	B	C	A	A

2.4 The Hare Method

We introduced preference lists as a way of representing a voter's thoughts about the various candidates; they were not actual, physical lists. However, a number of methods have been devised that require a voter to present a preference list. These methods are known collectively as *preferential voting*. Some methods require the voter to list all candidates; others allow a partial list. Preferential voting is most useful in its more general form, for situations where several representatives are to be elected at once (see Chap. 5). In this section we look at the simpler version.

The *Hare method* or *alternative vote* system was invented by the English lawyer Sir Thomas Hare in 1859. In 1871, William Robert Ware proposed what is essentially the same idea, so the name "Ware's method" is sometimes used. Other names are "instant runoff voting" and "preferential voting" (although the name "preferential voting" is also used for other preferential systems).

The Hare method *requires* each voter to provide a preference list at the election. This list is called a *ballot*. All candidates are to be listed, and no ties are allowed. The candidate with the fewest first place votes is eliminated. Then the votes are tabulated again as if there were one fewer candidate, and again the one with the fewest first place votes in this new election is eliminated. When only two remain, the winner is decided by a majority vote. We shall assume that each preference list contains all candidates—what we shall call a *complete preference list*, but in some places the Hare system has been modified so that a voter lists only those candidates of whom she/he approves.

The Hare system can be used to give an ordered list of all candidates in terms of the electorate's preference. At any stage, when a candidate is deleted, that candidate goes below all candidates who have not yet been deleted; if two are deleted simultaneously they will have the same number of first-place votes at that time, and are considered tied.

Sample Problem 2.2 *Suppose there are four candidates for a position, and 24 voters whose preference profile is:*

5	7	4	3	3	2
A	B	A	C	D	D
C	D	D	D	C	C
D	A	C	B	A	B
B	C	B	A	B	A

Who would win using the following electoral systems?

 (i) The plurality method.

 (ii) The runoff method.

(iii) The Hare method.

Solution.

 (i) The votes for A, B, C and D are 9, 7, 3 and 5 respectively, so A would win under plurality voting.

 (ii) Under the runoff method there is a tie. A and B are retained, and the new preference profile is:

5	7	4	3	3	2
A	B	A	B	A	B
B	A	B	A	B	A

,

giving 12 votes to each candidate.

(iii) In the Hare method we first eliminate C, obtaining

5	7	4	3	3	2
A	B	A	D	D	D
B	D	D	B	A	B
D	A	B	A	B	A

Now we eliminate B:

5	7	4	3	3	2
A	D	A	D	D	D
D	A	D	A	A	A

So D wins $15 - 9$.

Practice Exercise. Repeat the above question for the initial preference profile

6	7	7	7	2	7	5	2
A	B	A	C	D	D	B	D
C	D	D	D	C	C	D	B
D	A	C	A	A	B	C	C
B	C	B	B	B	A	A	A

As in all voting systems, it is possible that two candidates will receive the same number of first-place votes. If it happens that there is a tie between the two candidates with the smallest number of votes, both will be eliminated. This can lead to a candidate being elected without ever achieving more than the half the votes. For example, if the profile is

5	4	4
A	B	C
B	C	B
C	A	A

then both B and C will be eliminated, and A will win, although a majority prefer both B and C to A. However, such ties are extremely unlikely in political elections with typical numbers of voters.

As we said, one modification of the Hare method is to allow voters to cast votes only for the candidates of whom they approve. If there are five candidates, and you think A is best, B second, C third, but do not think either D or E is worthy of election, you simply vote A, B, C. If A, B and C are all eliminated, your vote is deleted, and the total number of votes cast is reduced accordingly. We shall refer to these generalized Hare systems as *instant runoff systems*.

In the above Sample Problem, suppose the seven voters represented by the second column all decided they did not wish to see A or D elected, while those represented by the fifth column did not like C. Then the preference profile is:

5	7	4	3	3	2
A	B	A	C	D	D
C	C	D	D	A	C
D		C	B	B	B
B		B	A		A

Again we first eliminate C,

5	7	4	3	3	2
A	B	A	D	D	D
B		D	B	A	B
D		B	A	B	A

Now we eliminate B:

5	4	3	3	2
A	A	D	D	D
D	D	A	A	A

and A wins 9–8.

Although nine votes is not a majority of the original 24 voters, it may be counted as a majority for this purpose. In some cases a *quota* is declared: a number is decided, and if neither of the last two candidates remaining achieves that many votes then the election is declared void. So, if there was a quota of 10 votes in the above example, the election would have to be held again.

A complication that can occur in real life is that a voter might not prefer one candidate over another: ties could occur in the voter's preferences. In practice, electoral systems that use preference lists do not allow ties, so that the voter must make a (possibly arbitrary) choice between the tied candidates. For simplicity, we shall assume that there are no ties in preference profiles.

Preferential voting is designed to avoid problems when there are three political parties all of which receive a substantial part of the vote. For example, suppose A and B are left-wing candidates and C is right-wing, and suppose the preference profile is

30	30	40
A	B	C
B	A	A
C	C	B

(where the numbers represent percentages). A majority, 60 %, of voters prefer the left-wing candidates, and would not wish to see C elected, but C would win under plurality. A would win under the preferential system.

The Hare system was introduced nationally in Australia in 1918. Up until then, the two major parties were the right-wing Liberal Party and the left-wing Labor Party. A new right-wing party, the Country Party, was formed to represent the interests of small farmers. The Country Party split the right wing vote in some country areas, allowing the Labor Party to win elections where the majority of voters would have preferred either of the other two parties to Labor.

The Hare system meant that voters could vote for their preferred candidate, but give second preference to a candidate with similar views. If the first choice was unsuccessful, the vote would go to the second choice. This the problem of split votes, and meant that voters' views were represented more accurately. The Hare method or something very similar is still used in all Australian state elections, and in the federal elections.

2.5 The Coombs Rule

The Coombs rule was proposed by American psychologist Clyde Coombs ([12], 397–399) as "an alternative to the Hare system." Coombs was primarily concerned with the analysis of psychological data, preferences, and such; we shall look at his method as an electoral scheme, but it will be more relevant later in our chapter on group preferences and committees.

Essentially, Coombs proposed a method for situations where every voter produces a complete preference list. If no candidate receives an absolute majority, the candidate with the greatest number of *last*-place votes is eliminated, the preference profile is updated, and the method is applied to the new profile.

Sometimes the Hare method and the Coombs rule give the same result. For example, in Sample Problem 2.2, first B is eliminated, with 12 last-place votes. The revised preference profile is

5	7	4	3	3	2
A	D	A	C	D	D
C	A	D	D	C	C
D	C	C	A	A	A

C is next to be eliminated, with 11 last-place votes. The final profile is the same as it was for the Hare method, so again D wins. However, as the next example illustrates, the two methods do not always give the same result.

Sample Problem 2.3 *Suppose there are four candidates for a position, and there are 33 voters whose preference profile is:*

10	8	7	6
A	B	C	D
D	D	B	C
B	C	D	B
C	A	A	A

Who would win using the following electoral systems?

(i) *The plurality method.*

(ii) *The runoff method.*

(iii) *The Hare method.*

(iv) *The Coombs rule.*

Solution.

(i) A is the plurality winner with 10 votes.

(ii) Under the runoff method, the two candidates remaining are A and B, and B wins $21 - 10$.

(iii) Under the Hare method, D is eliminated, producing the profile

10	8	13
A	B	C
B	C	B
C	A	A

Now B is eliminated, and C wins $21 - 10$.

(iv) Under Coombs, A is the first candidate eliminated, leaving profile

10	8	7	6
D	B	C	D
B	D	B	C
C	C	D	B

Next C, with 18 last places, is eliminated. D beats B $16 - 15$.
So the four methods produce four different winners.

2.6 Point Methods

Pointscore methods have often been used in sporting contests. For example, they are commonly used in track meets and in motor racing. When the Olympic Games are being held, many newspapers publish informal medal tallies to rank the performance of the competing nations—the usual method is to allocate three points for a gold medal, two for a silver and one for a bronze, and then add.

In general, a fixed number of points are given for first, second, and so on. The points are totalled, and the candidate with the most points wins. If there are n competitors, a common scheme is to allocate n points to first, $n - 1$ to second, ..., or equivalently $n - 1$ to first, $n - 2$ to second, ... This case, where the points go in uniform steps, is called a *Borda count*.

One often sees scales like 5, 3, 2, 1, where the winner gets a bonus, or 3, 2, 1, 0, 0, ... (that is, all below a certain point are equal). Sometimes more complicated schemes are used; for example, in the Indy Racing League, the following system has been used:

1st gets	20	5th gets	10	9th gets	4
2nd gets	16	6th gets	8	10th gets	3
3rd gets	14	7th gets	6	11th gets	2
4th gets	12	8th gets	5	12th gets	1

Fastest qualifier gets one point.
Leader of most laps gets one point.

Pointscore methods are occasionally employed for elections, most often for small examples such as selection of the best applicant for a job.

Sometimes the result depends on the point scheme chosen.

Sample Problem 2.4 *What is the result of an election with preference table*

5	7	4	3
A	B	A	C
C	C	C	B
B	A	B	A

if a 3, 2, 1 count is used? What is the result if a 4, 2, 1 count is used?

Solution. With a 3, 2, 1 count the totals are $A : 37, B : 36, C : 41$, so C wins. With a 4, 2, 1 count the totals are $A : 46, B : 43, C : 44$, and A wins.

Practice Exercise. Repeat the question for the preference table

7	3	2	1
A	B	B	C
B	A	C	B
C	C	A	A

2.7 Electing the American President

The President of the United States is not elected by the people. Rather, he or she is elected by the members of the Electoral College (the "electors"), who in turn are appointed using the popular vote. Each state is apportioned the same number of electors as the number of members of Congress to which the state is entitled, and District of Columbia receives the same number of electors as the least populous state, currently three. For example, Illinois has 18 members in the House of Representatives and 2 senators, so it has 20 electors; Indiana has 11. In total, there are currently 538 electors, corresponding to the 435 members of the House of Representatives, 100 senators, and the 3 additional electors from the District of Columbia.

The electors pledge their support for a presidential nominee. In most states, all the electors are pledged to the presidential candidate who wins the most votes in the state; in 2012, all 20 Illinois electors were pledged to Obama, and all 11 Indiana electors were pledged to Romney. In Maine and Nebraska, an elector is selected for each congressional district, pledged to the candidate representing the party that received the most votes, and two more are allocated to the party that received the most votes in the state overall. Each elector then casts one vote for President and another vote for Vice President. While the electors are not required by federal law to honor their pledge, there have been very few occasions when an elector voted otherwise.

If no candidate receives a majority of the electoral college votes for President, then the House of Representatives selects the President, with each state having precisely one vote. If no candidate receives a majority for Vice President, then the Senate selects the Vice President, with each Senator having one vote.

Candidates for elector are nominated by their state political parties in the months prior to Election Day. In some states, the electors are nominated in primaries, the same way that other candidates are nominated. In other states, electors are nominated in party conventions. In Pennsylvania, the campaign committee of each presidential candidate names their candidates for elector.

Opinions on the Electoral College system vary. The system can be seen to favor smaller states, because those states receive more electors per capita than larger states. For example, California had a population of just over 38 million in June 2012, and with 55 electoral votes it has approximately one elector for every 720,000 citizens; Indiana has approximately one per 595,000; and Wyoming has one per 192,000. On the other hand, the United States is a union of states, so some argue that different states should have equally many electors. For more information on the Electoral College, and in particular on the various opinions, see [34].

In the 2000 election, Al Gore received over 500,000 more individual votes than George Bush, about 5 % of the total number of voters. But Bush received the majority in 29 states plus the District of Columbia, while Gore won 21 states. Bush won 271 electoral votes to Gore's 266. The election was only decided after the Supreme Court voted to block recounts in Florida. The final tally in Florida showed Bush winning that state by 537 votes, less than 0.01 % of the Florida votes, so Bush received all 25 Florida electors. If the Florida decision had been different, Gore would have won by 291 electoral college votes to Bush's 256.

Exercises 2

1. How many votes are needed for a majority winner if there are 55 voters?

2. In how many ways can a voter rank six candidates assuming ties are not allowed?

3. Twenty-eight electors vote between candidates A, B and C. Their votes are 4 for A, 15 for B and 9 for C. What is the result under the majority method? What is the result under the plurality method?

4. Eighty-five electors vote between candidates A, B and C. There were 30 votes for A and 33 vote for B. How many votes did C receive? What is the result under the majority method? What is the result under the plurality method?

5. Students in the University games club are voting for a Club president. There are three candidates, Smith, Jones, and Brown. The preference table is

25	27	14	22	35
S	S	J	J	B
J	B	S	B	J
B	J	B	S	S

   (i) How many students voted?

   (ii) How many first place votes did each candidate receive?

   (iii) Who, if anybody, would win under the plurality method?

   (iv) Who, if anybody, would win under the majority method?

6. At the Academy Awards there are three nominees for Best Actor: Arthur Andrews, Bob Brown and Clive Carter. The preference table is

123	101	442	212	310
A	A	B	C	C
B	C	A	A	B
C	B	C	B	A

   (i) How many actors voted?

   (ii) How many first place votes did each candidate receive?

   (iii) Who, if anybody, would win under the plurality method?

   (iv) Who, if anybody, would win under the majority method?

   (v) Who would win under the runoff method?

7. In addition to plurality and the runoff method, two other techniques have been devised for cases when there is no majority winner. Both assume the full preference lists are known.

(a) The winner is the candidate with the fewest last-place votes.

(b) A runoff is held between the two candidates with the fewest last-place votes.

What are the results of using these two methods:

(i) Using the data of Exercise 5?

(ii) Using the data of Exercise 6?

8. Twenty-one electors must choose between five candidates: V, W, X, Y and Z. Their preference rankings are:

4	3	6	3	2	3
V	V	X	Y	Y	W
W	Y	Z	X	X	Z
X	W	Y	W	W	V
Y	X	V	Z	V	X
Z	Z	W	V	Z	Y

If there is no majority winner, all candidates with fewer than 20 % of the first-place votes (that is, those with fewer than 5) will be eliminated, and the preferences adjusted accordingly. If there is still no winner, a runoff is held. Which candidates are eliminated? What is the final result?

9. Given the following preference table, who would win under plurality voting? Who would win in a runoff?

6	3	4	2	1
A	C	C	B	E
E	B	D	A	A
B	E	A	C	B
C	D	E	D	C
D	A	B	E	D

10. A club with 36 members wishes to elect its president from four candidates, A, B, C and D. The preference profile is

16	10	8	2
A	B	C	B
B	A	B	A
C	D	A	C
D	C	D	D

  (i) Who would be elected if the club used plurality voting?

  (ii) Who would be elected if the club used the (3, 2, 1, 0) Borda count?

  (iii) Who would be elected if the club used a modified Borda count with scores (5, 2, 1, 0)?

11. Eighteen delegates must elect one of four candidates, A, B, C and D. The preference profile is

8	6	4
A	B	C
B	D	D
C	A	A
D	C	B

  (i) Who would be elected under the Hare method?

  (ii) Who would be elected under the Coombs rule?

  (iii) Who would be elected if the delegates used a modified Borda count with scores (2, 1, 0, 0)?

12. Fifty voters are to choose one of five candidates. Their preference profile is

20	10	14	6
A	B	B	C
C	A	A	D
E	C	D	B
B	D	C	A
D	E	E	E

What is the result under the following methods?

  (i) Plurality.

  (ii) Runoff.

  (iii) The Hare method.

  (iv) A modified Borda count with scores (5, 3, 2, 1, 0).

  (v) The Coombs rule.

13. Fifteen committee members are to choose a new treasurer from four candidates, A, B, C and D. Their preference profile is

7	5	3
A	C	D
B	B	C
D	A	B
C	D	A

What is the result under:

(i) Plurality?

(ii) The Hare method?

(iii) The Coombs rule?

14. One hundred voters choose between four candidates, A, B, C, D. Their preference profile is

40	32	10	18
A	B	C	D
C	C	D	C
B	A	A	B
D	D	B	A

(i) What is the result if the Hare method is used?

(ii) What is the result if the Coombs rule is used?

15. Suppose there are three candidates in an election, where one candidate is to be elected. Is there any difference whether the runoff method or the Hare method is used?

16. Twenty-five electors vote for three candidates, resulting in the preference table

8	6	7	4
X	Y	Z	Y
Z	Z	X	X
Y	X	Y	Z

It is decided to use a scoring system where first place gets n points, second gets 2 and third gets 1, where n is some whole number greater than 2. For what ranges of values is X winner? What is the range for Y? For Z? Does it ever happen that there is no result?

17. Explain why majority rule is a reasonable electoral method in a country with only two political parties, but is not good in a country with four major political parties. (Why is the word "major" important here?)

18. When deciding an election, is it necessary to know the number of voters associated with each preference list, or is it sufficient to know the percentage of voters?

19. Construct a preference profile for four voters and four candidates, such that three voters prefer W to X, three prefer X to Y, three prefer Y to Z, and three prefer Z to W.

Chapter 3
Simple Elections II: Condorcet's Method

We have already seen that, when there is no majority, different sensible-sounding electoral methods may produce different results. In 1785 Marie Jean Antoine Nicolas Caritat, Marquis de Condorcet, a French mathematician and political theorist, proposed a technique involving multiple use of runoff elections. (A similar idea was proposed by Ramon Llull as long ago as 1299; see for example [8].) Condorcet's work appeared in an essay entitled *Essai sur l'Application de l'Analyse à la probabilité des décisions rendues à la pluralité des voix* (Essay on the Application of Analysis to the Probability of Majority Decisions) [11]. This work also described several other results, including Condorcet's paradox, which shows that majority preferences become intransitive with three or more candidates.

3.1 The Condorcet Method

Suppose we simultaneously conduct all the "runoff" elections among our candidates. For example, in the election discussed in Sample Problem 2.2, there are six runoffs: A versus B, A versus C, A versus D, B versus C, B versus D and C versus D. If any one candidate wins all his/her runoffs, then surely you would consider that person a winner. We shall call such a candidate a *Condorcet winner.*.

In Sample Problem 2.1, we find:
B beats A 10–9,
A beats C 16–3,
B beats C 12–7,

© Springer International Publishing Switzerland 2014
W.D. Wallis, *The Mathematics of Elections and Voting*,
DOI 10.1007/978-3-319-09810-4_3

so B is a Condorcet winner. In Sample Problem 2.2,

A and B tie, 12–12,
A beats C 16–8,
D beats A 15–9,
C beats B 17–7,
D beats B 17–7,
D beats C 16–8.

So D is a Condorcet winner, even though A would win under plurality and the runoff method results in a tie between A and B.

But Condorcet's method does not yield a result in every set of preferences. In fact, Condorcet pointed out in the essay [11] that it is possible for a certain electorate to express a preference for A over B, a preference for B over C, and a preference for C over A from the same set of votes. Here is a simple example: given the preferences

5	4	3
A	B	C
B	C	A
C	A	B

A beats B 8–4, B beats C 9–3 and C beats A 7–5, so there is no Condorcet winner.

3.2 Condorcet's Extended Method

In elections with several candidates, it is very common to have no Condorcet winner, even when there are no ties. This is a serious fault in the Condorcet method.

Condorcet's own solution to this problem is as follows. We shall construct an ordered list of the candidates. Look at all the runoffs and find out which candidate won with the biggest majority. Looking at Sample Problem 2.1 again, the biggest majority was A beat C 16–3. We'll denote this $A \rightarrow C$. Then look for the second-biggest, then the third-biggest, and so on, and make a list:

$$A \rightarrow C(16 - 3), B \rightarrow C(12 - 7), B \rightarrow A(10 - 9).$$

Now go through this list and construct a preference order of the candidates. At each step, if $X \rightarrow Y$, then X precedes Y in the preference list, *unless* Y already precedes X in the list. In our example, we must have A before C, B before C and B before A. The list is BAC and clearly B is the winner.

We shall refer to this solution as *Condorcet's extended method*, to distinguish it from the case where there is a Condorcet winner under the original method. Note that, if there is a Condorcet winner, the same candidate also wins under the extended method.

Let us apply this to the above example

5	4	3
A	B	C
B	C	A
C	A	B

which has no Condorcet winner. We have, with the larger majorities preceding smaller ones,

$$B \to C(9 - 3), A \to B(8 - 4), C \to A(7 - 5).$$

From $B \to C$ and $A \to B$ we get the list ABC. Next we see $C \to A$, but A already precedes C, so this result is ignored. The final list is ABC, and A is elected, even though a majority of voters would prefer C to A.

Sample Problem 3.1 *Consider the election with preference profile:*

7	5	3
A	B	C
B	C	A
C	A	B

Who would win under the Hare method? Is there a Condorcet winner? Who wins under Condorcet's solution method?

Solution. The votes for A, B and C are 7, 5 and 3 respectively. Under the Hare method, C is eliminated. The new preference profile is:

7	5	3
A	B	A
B	A	B

that is,

10	5
A	B
B	A

So A wins 10−5. Looking at all three runoffs, we see that A beats B 10−5, B beats C 12−3 and C beats A 8−7, so there is no Condorcet winner. For Condorcet's solution, we see

$$B \to C(12 - 3), A \to B(10 - 5), C \to A(7 - 5).$$

The first two yield the list ABC and the last result is ignored, so A is elected.

Practice Exercise. Consider the election with preference profile:

6	4	3
A	B	C
B	C	A
C	A	B

Who would win under the Hare method? Is there a Condorcet winner? Who wins under Condorcet's solution method?

3.3 Condorcet Winner Criterion

We have seen that not every preference profile leads to a Condorcet winner. However, in those cases where there is a Condorcet winner, it seems reasonable to expect that the Condorcet winner would also be the winner under other methods. But this is not always true.

An electoral system is said to satisfy the *Condorcet winner criterion* if, whenever there is a Condorcet winner, then the electoral system in question will always choose the Condorcet winner.

The Hare system. The Hare system does not satisfy the Condorcet winner criterion. As an example, suppose 100 voters have the preference profile

40	21	39
A	B	C
B	C	B
C	A	A

Then 60 voters prefer B to A and 61 prefer B to C, so B is a Condorcet winner. However, B received the smallest number of first-place votes, and is the first candidate eliminated. The resulting preference profile is

40	21	39
A	C	C
C	A	A

and C wins.

Borda count. The Borda count does not satisfy the Condorcet winner criterion. To see this, consider the following profile. There are 3 candidates, A, B and C, and the 11 voters have preference profiles

6	4	1
A	B	C
B	C	A
C	A	B

Seven of the eleven voters prefer A to B, while six prefer A to C. So A is a Condorcet winner. However, in a (2,1,0) Borda count, B scores 14, A scores 13, and C scores 6. So B wins.

Plurality voting. The example used in discussing the Hare system can also be used to show that plurality voting does not satisfy the Condorcet winner criterion. As we saw, B is a Condorcet winner, but the winner under plurality voting would be A (and not C, who won under the Hare system).

Majority voting. It is clear that if there is a winner under the majority system, that candidate will have more than half the first-place votes, and consequently will have more than half the votes when compared to any competitor. So the winner is also a Condorcet winner. However, it is quite possible that there will be no majority winner, even when there is a Condorcet winner. (Again, the preference profile for the Hare system provides an example.) So majority elections do not satisfy the Condorcet winner criterion.

3.4 Condorcet Loser Criterion

A *Condorcet loser* in an election is a candidate who would lose in a runoff against each other candidate. Not every election with three or more candidates has a Condorcet loser. For example, in the election shown in Sample Problem 3.1, there is no Condorcet loser.

A voting system satisfies the *Condorcet loser criterion* if it can never happen that a Condorcet loser wins the election. Plurality voting does not satisfy this condition, but several methods, including the Hare system, do.

Plurality voting. Consider the following simple example, with three candidates.

2	7	5	6
A	A	B	C
B	C	C	B
C	B	A	A

A is a Condorcet loser—both B and C beat A 11:9. However, A has the largest number of first-place votes.

The Hare system. Suppose A is the winner of a Hare election. After various candidates have been eliminated, A is one of the remaining candidates, and receives a majority of the votes when compared to the other remaining candidate or candidates. Therefore a majority of voters place A above the other candidates that survived to the last round. It follows that A would win a runoff against any of those candidates, and therefore is not a Condorcet loser.

3.5 The Bucklin Method

Another method that Condorcet proposed, in 1793, was a technique to avoid the problems when there is no majority winner. The system was revived by James W. Bucklin of Grand Junction, Colorado in the early 1900s, and is now called the

Bucklin Method or the Grand Junction System. It was used in a number of American states in the early twentieth century. For a partial history of its use in Minnesota, see [27].

In the Bucklin method, each voter submits a preference list. The first preferences are tallied, and if any candidate receives a majority, she or he is elected. Let us refer to the number of votes required for a majority the *quota*.

Suppose no candidate achieves a majority. Then the second votes are added to the tally. In effect, each elector votes for two candidates. If one candidate receives the quota or higher, that candidate is elected. As the number of votes has been doubled, it is possible that more than one candidate will achieve a majority, and in that case the one with the bigger total wins.

As an example, consider the following preference profile.

8	6	4	2
A	B	C	D
B	C	D	C
C	D	B	B
D	A	A	A

There are 20 voters, so the quota is 11. Initially, the votes are $8, 6, 4, 2$ for A, B, C, D respectively, so there is no winner. After the second preferences are added, the votes are $8, 14, 12, 6$, so both B and C achieve the quota; B's total is larger, so B is elected.

Of course, it is also possible that there will still be no winner after the second tally. This is particularly true when there are a large number of candidates. In that case, third preferences are added, and so on. An example:

6	4	6	4
A	B	C	D
B	C	D	A
D	D	B	C
C	A	A	B

Again the quota is 11. After the first count, the totals are $6, 4, 6, 4$ for A, B, C, D respectively, so there is no winner. After the second count, the totals are $10, 10, 10, 10$, so again there is no winner. The third round totals are $10, 16, 14, 20$, so D is elected.

As in any system, it is possible that the result will be a tie. The usual response is to hold another election, but in the Bucklin system there is another possibility. If the quota has been reached, and two candidates are tied for first place, add another round of votes; if one of the tied candidates receives more than the other, that one is a winner.

Sample Problem 3.2 *What is the result of an election with preference profile*

5	4	3	2	2	2
A	B	C	C	D	D
B	A	D	A	B	C
C	C	A	B	A	B
D	D	B	D	C	A

if the Bucklin method is used?

Solution. There are 18 voters, so the quota is 10. The initial totals are 5, 4, 5, 4 for A, B, C, D respectively, so there is no winner after the first round. After second preferences are added, the totals in the same order are 11, 11, 7, 7, so both A and B have achieved enough votes, but they are tied. So a third round is calculated. The new totals are 16, 15, 16, 7, so A beats B and is elected. Round two was when the quota was achieved, so only those who were tied for first place at that time are considered, even though C now has as many votes as A.

There is a modification available, which we shall call the *modified Bucklin method*. Voters only list those candidates of whom they approve. Then some columns of the profile will contain some blank cells at the bottom.

3.6 Condorcet Voting Systems

We have seen that many methods do not satisfy the Condorcet winner criterion. Because of this, several methods that *do* satisfy the criterion have been invented. These methods are often called *Condorcet voting systems*.

One very simple system was proposed by the Scottish economist Duncan Black [4] in 1958. He proposed using both Condorcet's method and some sort of Borda count. If there is a Condorcet winner, then that candidate wins the election. If not, the Borda count is calculated, and the Borda count winner is elected. This obviously satisfies the Condorcet winner criterion.

An earlier method combining Condorcet's methods and Borda count was proposed by E. J. Nanson, an English mathematician who migrated to Australia in the late nineteenth century. His method is as follows. First, conduct a Borda count. Then eliminate any candidate who receives less than the average score. (For example, if there are n voters and five candidates, and scores $4 - 3 - 2 - 1 - 0$ are used, the average score will be 2 per ballot, and the average overall score will be $2n$.) Then recalculate the scores of the remaining candidates as though they had been the only candidates. Eliminate the candidate with the lowest score. Repeat this process, until there is only one candidate, or a collection of tied candidates, remaining. This process will always select a Condorcet winner, if there is one. (See Exercise 3.14.)

This method was modified by the Australian astronomer J. M. Baldwin, who proposed eliminating only the lowest scorer at each stage. For more on these methods, see [19].

A. H. Copeland proposed the following method in a seminar at the University of Michigan in 1951. For a given candidate, consider all pairwise comparisons between that candidate and others. If the candidate would win in x cases and lose in y cases, then her score is $x - y$. (Comparisons resulting in a tie are ignored.) We shall examine this method in Exercise 3.12.

Sample Problem 3.3 *What is the result of an election with preference profile*

16	10	10	4
A	C	B	C
B	A	C	B
C	D	A	D
D	B	D	A

if Nanson's method is used?

Solution. There are 40 voters. Assuming a 3-2-1-0 Borda count is used, the Borda counts are $A - 78$, $C - 78$, $B - 70$, $D - 14$, The average score is 1.5 per vote, for a total of 60. So D is eliminated. The new profile is

16	10	10	4
A	C	B	C
B	A	C	B
C	B	A	A

and the counts are $A - 42$, $B - 40$, $C - 38$, and the average is 40. Now C is eliminated. Finally, A beats B 26–14 in a head-to-head vote.

3.7 Sequential Pairwise Voting

Condorcet's method essentially looks at all comparisons of two candidates. Another method that involves breaking an election into a number of smaller, two-candidate, elections is *sequential pairwise voting*, In this, several candidates are paired in successive runoff elections. There is an *agenda* (an ordered list of candidates). For example, if the agenda is $A, B, C, D, \ldots$ then the elections proceed as follows:

1. A against B

2. Winner of AB against C

3. That winner against D

    . . .

Position in the agenda is very important. To see this, consider a four-candidate election with agenda A, B, C, D, in which all four candidates are equally likely to win. If repeated trials are made then we would expect the following results:

- A wins first runoff in half the cases
- A wins second runoff in half those cases—a quarter overall
- A wins the third runoff in half those cases—one-eighth overall.

So A has a 1 in 8 chance of winning. B also has a has 1 in 8 chance. However, C has a 1 in 4 chance, and D has a 1 in 2 chance. In this case being later in the list is very beneficial.

More generally, consider an election with four candidates and only three voters. One possible preference profile is:

1	1	1
A	B	C
B	C	D
C	D	A
D	A	B

There is an agenda to favor every candidate. For example, if the agenda is B, C, D, A, then B beats C 2–1, B beats D 2–1, and A beats B 2–1, so A wins. Using A, C, B, D we see C beat A 2–1, then B beats C and B beats D, a victory for B. If you wish C to win, use the agenda A, B, C, D, while B, C, A, D gives the election to D.

Rather than elections, the sequential pairwise model is often used for sporting tournaments (the result of match is used instead of the result of a runoff election). One often sees playoff rules like:

(i) Second and third placegetters in preliminary competition play each other ("the playoff");

(ii) The winner of playoff meets the leader from the preliminaries.

In this case it is reasonable that the preliminary leader should get an advantage. However, when the model is used in voting situations, it is very subject to manipulation.

Exercises 3

1. Consider the following statements.

   (i) If X would win a plurality election, then X would win under Condorcet.

   (ii) If X would win a majority election, then X would win under Condorcet.

   Is either of these statements always true?

2. Consider the following preference table. In Chap. 2, Exercise 10, we asked, who would win under plurality voting? Who would win in a runoff? The answers were C and A respectively.

6	3	4	2	1
A	C	C	B	E
E	B	D	A	A
B	E	A	C	B
C	D	E	D	C
D	A	B	E	D

   Is there a Condorcet winner in this example? If not, who would win under Condorcet's extended method? Who would win under Bucklin's method?

3. A club with 36 members wishes to elect its president from four candidates, A, B, C and D. The preference profile is

16	10	8	2
A	B	C	B
B	A	B	A
C	D	A	C
D	C	D	D

   Is there a Condorcet winner? If not, who would win under Condorcet's extended method? Who would win under Bucklin's method?

4. Eighteen delegates must elect one of four candidates, A, B, C and D. The preference profile is

8	6	4
A	B	C
B	D	D
C	A	A
D	C	B

   Is there a Condorcet winner? If not, who would win under Condorcet's extended method?

5. Here is the preference profile for an election with three candidates, A, B, C. Is there a Condorcet winner? If not, who would win under Condorcet's extended method? Who would win under Bucklin's method?

8	5	6	4
A	C	B	B
C	A	C	A
B	B	A	C

6. Fifty voters are to choose one of five candidates. Their preference profile is

20	10	14	6
A	B	B	C
C	A	A	D
E	C	D	B
B	D	C	A
D	E	E	E

Is there a Condorcet winner? If not, who would win under Condorcet's extended method? Who would win under Bucklin's method?

7. Fifteen committee members are to choose a new treasurer from four candidates, A, B, C and D. Their preference profile is

7	5	3
A	C	D
B	B	C
D	A	B
C	D	A

Is there a Condorcet winner? If not, what is the result under the extended Condorcet method?

8. One hundred voters choose between four candidates, A, B, C and D. Their preference profile is

40	32	10	18
A	B	C	D
C	C	D	C
B	A	A	B
D	D	B	A

If the Hare method is used, the result is a tie between A and B. Is there a Condorcet winner? If not, what is the result under the extended Condorcet method? Who would win under Bucklin's method?

9. Twenty voters choose between four candidates, A, B and C. Their preference profile is

5	5	1	3	2	4
A	A	B	B	C	C
B	C	A	C	A	B
C	B	C	A	B	A

Is there a Condorcet winner? If not, what is the result under the extended Condorcet method? Who would win under Bucklin's method?

10. Verify that if an election has a Condorcet winner, then the same candidate also wins under Condorcet's extended method.

11. Does the Bucklin method satisfy the Condorcet winner criterion?

12. Prove that the Copeland method satisfies both the Condorcet winner and Condorcet loser criteria. What is the shortcoming of this method?

13. Verify that the Black rule satisfies the Condorcet winner criterion.

14. Verify that Nanson's method will always select a Condorcet winner, if there is one.

15. Who will win in the election described in Exercise 7 under Black's rule? Who would win under Nanson's method?

16. Use the preference profile

5	4	4	3
W	W	X	X
X	X	Y	T
Z	T	T	Z
T	Z	X	X

to show that the Borda count and the Nanson method do not always give the same result.

17. Consider the preference profile

3	1	1	1	1	1	1
W	W	X	X	Y	Y	Z
Z	Y	Y	Z	X	X	X
Y	Z	Z	Y	Z	W	Y
X	X	W	W	W	Z	W

What is the result of an election using the following systems?

(i) Majority system.

(ii) Plurality system.

(iii) Borda count.

(iv) Hare system.

(v) Condorcet method.

(vi) Sequential pairwise method with agenda X, Y, Z, W.

(vii) Sequential pairwise method with agenda W, X, Z, Y.

(viii) Sequential pairwise method with agenda Y, Z, W, X.

18. Thirty board members must vote on five candidates: X, Y, Z, U and V. Their preference rankings are summarized in the table below. Find the winner using sequential pairwise voting with the agenda X, Y, Z, U, V.

12	10	8
X	Y	Z
U	U	U
Y	Z	X
Z	X	V
V	V	Y

19. An 18-member committee is to elect one of the four candidates Q, R, S, T. Their preference table is as shown. Which candidate wins under sequential pairwise voting with agenda (S, T, Q, R)?

4	6	5	3
Q	R	S	T
T	S	R	S
S	T	T	R
R	Q	Q	Q

20. Consider a sequential pairwise election with preferences

5	3	2	1	8
X	X	Y	Z	T
Y	T	Z	Y	Y
T	Y	X	X	X
Z	Z	T	T	Z

    (i) Who will win under agenda X, Y, Z, T?

    (ii) Who will win under agenda Y, T, Z, X?

21. A sequential pairwise election has preference profile

2	8	11	7
A	C	B	A
B	B	A	C
C	A	C	B

    (i) Who will win under agenda A, B, C?

    (ii) Who will win under agenda B, C, A?

    (iii) Who will win under agenda C, A, B?

22. A sequential pairwise election has preference profile

40	32	12	18
P	Q	R	S
Q	R	S	R
R	P	P	Q
S	S	Q	P

(i) Who will win under agenda P, Q, R, S?

(ii) Who will win under agenda S, R, Q, P?

(iii) Who will win under agenda S, P, R, Q?

(iv) Who will win under agenda S, Q, P, R?

23. Does the plurality runoff method satisfy the Condorcet winner criterion?

24. Does sequential pairwise voting satisfy the Condorcet winner criterion?

Chapter 4
Fair Elections; Polls; Amendments

We have seen several different electoral systems. Even in a small election, the same preferences can give rise to a different results depending on the system used. In this section we shall present an extreme case of how the method chosen might affect the outcome of an election in a realistic situation. This means that an electoral system could be chosen in order to favor one or another candidate.

There are also many instances in which voters could misrepresent their preferences, manipulating an election to give a certain result. This can be unintentional; when polls are held before an election, the results may convince some voters that their favorite candidate has no chance, and they may decide to vote for their second-favorite rather than risk their least preferred candidate gaining election. Or it can be intentional, as we shall see in our discussion of amendments.

4.1 Five Candidates, Six Methods, Six Results

Consider a political party convention at which six different voting schemes are adopted. Assume that there are 110 delegates to this national convention, at which five of the party members, denoted by A, B, C, D, and E, have been nominated as the party's presidential candidate. Each delegate must rank all five candidates according to his or her choice. Although there are $5! =: 5 \times 4 \times 3 \times 2 \times 1 = 120$ possible rankings, many fewer will appear in practice because electors typically split into blocs with similar rankings. Let's assume that our 110 delegates submit only six different preference lists, as indicated in the following preference profile:

© Springer International Publishing Switzerland 2014
W.D. Wallis, *The Mathematics of Elections and Voting*,
DOI 10.1007/978-3-319-09810-4_4

36	24	20	18	8	4
A	B	C	D	E	E
D	E	B	C	B	C
E	D	E	E	D	D
C	C	D	B	C	B
B	A	A	A	A	A

The 36 delegates who most favor nominee A rank D second, E third, C fourth, and B fifth. Although A has the most first-place votes, he is actually ranked last by the other 74 delegates. The 12 electors who most favor nominee E split into two subgroups of 8 and 4 because they differ between B and C on their second and fourth rankings.

We shall assume that our delegates must stick to these preference schedules throughout the following six voting agendas. That is, we will not allow any delegate to switch preference ordering in order to vote in a more strategic manner or because of new campaigning.

We report the results when six popular voting methods are used. There are six different results.

1. **Majority.** As one might expect with five candidates, there is no majority winner.

2. **Plurality.** If the party were to elect its candidate by a simple plurality, nominee A would win with 36 first-place votes, in spite of the fact that A was favored by less than one-third of the electorate and was ranked dead last by all other delegates.

3. **Runoff.** On the other hand, if the party decided that a runoff election should be held between the top two contenders (A and B), who together received a majority of the first-place votes in the initial plurality ballot, then candidate B outranks A on 74 of the 110 preference schedules and is declared the winner in the runoff.

4. **Hare Method.** Suppose the Hare method is used: a sequence of ballots is held, and at each stage the nominee with the fewest first-place votes is eliminated. The last to survive this process becomes the winning candidate. In our example E, with only 12 first-place votes, is eliminated in the first round. E can then be deleted from the preference profile, and all 110 delegates will vote again on successive votes. On the second ballot, the 12 delegates who most favored E earlier now vote for their second choices, that is, 8 for B and 4 for C; the number of first-place votes for the 4 remaining nominees is

$$A \quad B \quad C \quad D$$
$$36 \quad 32 \quad 24 \quad 18$$

Thus, D is eliminated. On the third ballot the 18 first-place votes for D are reassigned to C, their second choice, giving

$$A \quad B \quad C$$
$$36 \quad 32 \quad 42$$

Now B is eliminated. On the final round, 74 of the 110 delegates favor C over A, and therefore C wins.

5. **Borda count.** Given that they have the complete preference schedule for each delegate, the party might choose to use a straight Borda count to pick the winner. They assign five points to each first-place vote, four points for each second, three points for a third, two points for a fourth, and one point for a fifth. The scores are:

A: $254 = (5)(36) + (4)(0) + (3)(0) + (2)(0) + (1)(24 + 20 + 18 + 8 + 4)$

B: $312 = (5)(24) + (4)(20 + 8) + (3)(0) + (2)(18 + 4) + (1)(36)$

C: $324 = (5)(20) + (4)(18 + 4) + (3)(0) + (2)(36 + 24 + 8) + (1)(0)$

D: $382 = (5)(18) + (4)(36) + (3)(24 + 8 + 4) + (2)(20) + (1)(0)$

E: $378 = (5)(8 + 4) + (4)(24) + (3)(36 + 20 + 18) + (2)(0) + (1)(0)$

The highest total score of 382 is achieved by D, who then wins. A has the lowest score (254) and B the second lowest (312).

6. **Condorcet.** With five candidates, there is often no Condorcet winner. However, when we make the head-to-head comparisons, we see that E wins out over:

- A by a vote of 74–36
- B by a vote of 66–44
- C by a vote of 72–38
- D by a vote of 56–54

So there is a Condorcet winner, namely E.

In summary, our political party has employed six different common voting procedures and has come up with five different winning candidates. We see from this illustration that those with the power to select the voting method may well determine the outcome.

4.2 Manipulating the Vote

The term *strategic voting* means voting in a way that does not represent your actual preferences, in order to change the result of the election. We would call the resulting ballot *insincere.*

Suppose your favorite is candidate X. (We will call you an X *supporter.*) Then X would normally appear at the top of your preference list. But sometimes you can achieve X's election by voting for another candidate in first place! This is most common in runoff situations; you can ensure that your candidate does not have to face a difficult opponent. The following example illustrates this.

Sample Problem 4.1 *A runoff election has preference profile*

6	2	7	5	4
A	C	C	B	D
D	B	A	A	A
C	D	D	C	B
B	A	B	D	C

Show that the supporters of C can change the result so that their candidate wins, by the two voters in the second column changing their ballots by demoting *their candidate.*

Solution. Initially the first-place votes are A–6, B–5, C–9, D–4, so runoff election will be between A and C, and A wins 15–9. The revised profile is

6	2	7	5	4
A	B	C	B	D
D	C	A	A	A
C	D	D	C	B
B	A	B	D	C

,

the first-place votes are A–6, B–7, C–7, D–4, so the runoff election is between A and C, and C wins 13–11.

Even when you cannot ensure victory for your opponent, you may still be able to obtain a preferable result. For example, suppose you support candidate X; you think candidate Y is acceptable, but hate candidate Z. Even if insincere voting cannot ensure victory for candidate X, you may be able to swing the election to Y rather than Z.

Sample Problem 4.2 *An election with four candidates and seven voters is to be decided by the Hare system. The preference profile is*

2	1	2	1	2
B	B	D	C	C
A	D	C	A	D
D	A	B	B	A
C	C	A	D	B

Show that one of the two voters with preference list B, A, D, C can change the outcome to a more favorable one by insincere voting.

Solution. First consider the result of sincere voting. Initially A is eliminated, having no first-place votes:

2	1	2	1	2
B	B	D	C	C
D	D	C	B	D
C	C	B	D	B

Next D is eliminated, leaving

2	1	2	1	2
B	B	C	C	C
C	C	B	B	B

The winner is C.

Now suppose one voter changes his ballot from B, A, D, C to D, A, B, C. The profile is

1	1	1	2	1	2
B	D	B	D	C	C
A	A	D	C	A	D
D	B	A	B	B	A
C	C	C	A	D	B

Again A is first eliminated, leaving:

1	1	1	2	1	2
B	D	B	D	C	C
D	B	D	C	B	D
C	C	C	B	D	B

In the next round, B is eliminated.

1	1	1	2	1	2
D	D	D	D	C	C
C	C	C	C	D	D

The winner is D. This is a preferable outcome for the voter who switched.

Practice Exercise. Consider a Hare system election with preference profile

8	6	5
P	Q	R
Q	R	S
R	P	P
S	S	Q

Show that P would win this election. Show that if one of the six supporters of Q changes her vote, she could ensure that R wins, even though a majority of voters still prefer Q to R.

4.3 Polls

Of course, you do not always know exactly how the votes will go. Strategic voting is usually based on assumptions about the election. How do we arrive at these assumptions?

Often an election is preceded by an informal count, or poll. For example, as early as 2011 there were several polls for the 2012 Presidential election in the United States.

If a candidate does badly in polls, his or her supporters may change their votes. For example, consider an election that uses plurality voting. There are 390 voters and 3 candidates, A, B, C. The voters' actual preferences are

180	170	40
A	B	C
C	C	B
B	A	A

If the election were held immediately, A would win.

However, suppose a poll is held. As usual, the poll uses the same system (plurality voting). Say there are 39 voters in the poll, and their preferences are proportional to the overall preferences. The result will show A first (18 votes), with B a close second (17 votes), and C a distant third (with only 4 votes).

When the poll results are reported, the 40 voters who favor C may well reason, "C cannot win the election, so it would be preferable to elect the better of the other two." They all prefer B to A, so they would vote for B. Suppose 30 of the voters decided to change their votes. The vote will be A–180, B–200, C–10, and B will win the election.

But observe that C was a Condorcet winner under the original preferences:

B beats A 220–170
C beats B 220–170
C beats A 210–180

The type of behavior exhibited by C's followers is frequently observed after polls and leads to the

Poll Assumption:

   Voters whose favorite is not one of the two top candidates in a poll will adjust their preferences to vote for the one of the top two that they prefer.

For this reason, those who do badly in polls may drop out before the election. In the example, C might easily have dropped out. The Poll Assumption applies primarily to the last poll taken before an election, but candidates frequently drop after earlier polls, because they do not expect their showing to improve later.

In cases where the electors' preference list is involved, we assume the voters who adjust their preferences will do so by moving the new preferred candidate to the top of the list, but will not change their relative preferences among the other candidates.

Theorem 1. *If the poll assumption holds, and a poll is held prior to a plurality election, a Condorcet winner will win the election if and only if she is one of the top two candidates in the poll.*

Proof. Suppose the two candidates who topped the poll were A and B. Since only A and B retain first place positions in the final election preferences, the election is the same as a majority election between A and B. The relative positions of A and B are unchanged, so if A is a Condorcet winner then A beats B in the final election. A Condorcet winner who is not one of the top two will not be A or B, so cannot win.
□

In cases where the electors' preference list is involved, we assume the voters who adjust their preferences do so by moving the new preferred candidate to the top of the list, but will not change their relative preferences among the other candidates.

We say an election satisfies the *poll fairness criterion* if there is no case where the result is changed by a candidate who is not one of the top two drops out before the final vote; a system satisfies the poll fairness criterion if every instance satisfies the condition.

4.4 Sequential Pairwise Voting

As we noted in Chap. 3, sequential pairwise voting is very prone to manipulation. This can involve both the possibility of insincere voting and the order of candidates in the voting agenda.

As an example, suppose there are four political parties. Party A is extremely left-wing, party B is moderately left-wing, party C is moderately right-wing, and party D is extremely right-wing. As we would expect, each voter puts the two candidates with his or her political leaning first and second, the moderate on the other side third and extremist on the other side last. Most of the electorate turns out to be extremist in one way or the other, and the voters' preferences are

11	1	1	10
A	B	C	D
B	C	B	C
C	A	D	B
D	D	A	A

If the voting agenda is A, D, B, C then the results are that A beats D 12–11, then B beats A 12–11, and finally B beats C 12–11. However, suppose the voter represented by the third column changes her preference list to C, A, B, D. Then initially A beats D 13–10 and then A beats B 12–11. But then C beats A 12–11. By increasing her support of the extremist of the opposite political persuasion, the supporter of the moderate left-winger has achieved her preferred candidate's election.

However, this result completely depends on the voting agenda. If the agenda B, C, A, D had been chosen, and all electors voted sincerely, the reflection would still favor B: B beats C 12–11, B beats A 12–11, and B beats D 13–10. But if the supporter of C changes her preference list to C, A, B, D, as before, the result is that initially B beats C 12–11, as before, but then A beats B 12–11, and finally A beats D 13–10. The moderate right-winger has caused the extreme left-winger to be elected!

4.5 Amendments

We now consider an important example of manipulation that involves the introduction of sequential pairwise voting into what was originally a straightforward majority vote. Suppose three voters on City Council have to decide whether to add a new sales tax. Initially

- A prefers the tax
- B prefers the tax
- C prefers no tax

so a tax will be introduced.

However, let's assume A hates income taxes and will never vote for one. On the other hand, B prefers income tax to sales tax. Suppose C moves an amendment to change the tax to an income tax.

We now have:

Original motion: that a city sales tax of 5 % be introduced.

Amendment (moved by C): change "sales tax of 5 %" to "income tax of 2 %". (We shall assume the 2 % income tax will provide the same total as the 5 % sales tax.)

In the vote on the amendment, both B and C will vote in favor, with A against, so the amendment is carried. So the motion becomes: a city income tax of 2 % shall be introduced. In the vote on the new motion, both A and C are against, while B votes in favor; so the motion is lost and there is no tax.

A related technique can be applied in regular elections. Suppose a plurality election is to be held in a two-party system (for example, for the United States House or Senate). If a third candidate, who has similar views to one of the original two, were added to the ballot, and some of the electors vote for the new candidate, this could change the result. For example, say 51 % intend to vote for candidate A and 49 % for candidate B. If a third candidate with very similar views to A, say C, decides to run and gains 5 % of the vote, it could well be that B will still receive 49 %, A will receive only 46 %, and B will win. There have been cases where voters of one political persuasion will try to convince an independent to run whose views are opposite to their own. This is one of the reasons that some electoral systems have been changed to various kinds of preferential voting.

Exercises 4

1. What would be the result of the election described in Sect. 4.1 under the Coombs Rule?

2. What would be the result of the election described in Sect. 4.1 if the Bucklin method was used?

3. A committee of 11 members needs to elect one representative from four candidates. They plan to use a Borda count. The preferences are

4	2	2	3
A	A	B	B
C	B	C	D
D	C	A	C
B	D	D	A

   (i) Who would win the election if all electors vote sincerely?

   (ii) Can the voters in the third column vote insincerely so as to change the result in their favor? If so, how?

   (iii) Can the voters in the fourth column vote insincerely so as to change the result in their favor? If so, how?

   (iv) If the Hare method is used, can the voters in the third and fourth column change the result by voting insincerely?

4. Suppose a 30-voter election with the following preference profile is decided by the Hare system.

3	6	7	8	6
A	A	B	C	D
B	C	D	D	C
C	D	A	A	B
D	B	C	B	A

   (i) Who will win the election?

   (ii) Show that the seven supporters of B can achieve a preferred result if they exchange B and D on their ballots (that is, they vote as if their preference was D, B, A, C).

   (iii) Who will win a plurality election?

5. In the preceding exercise (Question 2), who would win under the Bucklin method, and who would place second? Can the second-place candidated vote insincerely so as to ensure their candidate wins?

6. An election has preferences

3	2	2	2
A	B	C	D
B	D	B	C
C	C	A	B
D	A	D	A

(i) Who will win the election under plurality?

(ii) Who will win under the Bucklin method?

(iii) Who will win under Borda count?

(iv) Show that the supporters of C can make sure their candidate wins under Borda count by insincerely voting preference (C, A, D, B).

7. Consider a runoff election with preferences

4	4	3	2	4
A	A	B	B	C
B	C	A	C	B
C	B	C	A	A

(i) Who wins the election?

(ii) Show that two supporters of A can change the result in favor of their candidate by changing their preferences from (A, C, B) to (C, A, B).

8. Consider the preferences

24	16	19	21	9	10
A	B	C	D	E	E
B	A	D	C	C	C
E	D	E	E	B	A
D	E	A	B	A	B
C	C	B	A	D	D

(i) Suppose everyone votes sincerely. Under the Hare system, who wins the election?

(ii) Show that, if one of the supporters of D votes insincerely by reversing her first two preferences, she achieves a preferable outcome.

9. An election with preference profile

8	4	3
A	B	C
B	C	B
C	A	A

is decided by Borda count. How many of A's supporters must change their ballots to (A, C, B) in order to make their candidate the winner?

10. A club with 46 members wishes to elect a president. The post is currently held by Y. The 46 members have preference profile

16	12	10	8
X	Y	Z	Y
Z	Z	X	X
Y	X	Y	Z

(i) The following electoral system is used: a plurality vote is held between the candidates other than the President; there is then a runoff election between the winner of that election and the current President. Show that X will win the election.

(ii) The eight voters whose votes form the right-hand column of the profile decide to change their votes in an attempt to make Y the winner. Can they do this?

11. An election is to be held under plurality voting. The preferences profile is

1	3	2
A	C	B
C	B	A
B	A	C

Can the supporter of A (first column) achieve a preferable result by insincere voting?

12. A plurality-vote election has preference profile

8	6	4	2
X	Y	Z	T
T	T	T	Z
Y	X	Y	Y
Z	Z	X	X

(i) Is there a Condorcet winner? If so, who?

(ii) Suppose a poll is held and those who place third and fourth drop out. Who will win the final election?

(iii) Suppose a poll is held but only the candidate who placed fourth drops out. Who will win the final election?

13. A hiring committee uses the Hare system to select a new foreman. The preference profile is

5	4	3
A	B	C
D	A	D
B	D	B
C	C	A

(i) Who would win this election?

(ii) Who would win if C drops out before the election?

(iii) Who would win if A drops out before the election?

14. Fifteen club members are voting to elect a president, using a Borda count. The preference profile is

6	5	4
A	B	C
C	D	B
B	A	A
D	C	E
E	E	D

A poll is conducted; both C and D realize that they are unlikely to win.

(i) Who would win this election?

(ii) Who would win if C drops out before the election?

(iii) Who would win if D drops out before the election?

15. Consider the sequential pairwise election with profile

11	1	1	10
A	B	C	D
B	C	B	C
C	A	D	B
D	D	A	A

that we discussed in Sect. 4.4. Is there any voting agenda that changes the result, provided all electors vote sincerely?

Chapter 5
Arrow's Theorem and the Gibbard-Satterthwaite Theorem

In many voting systems, each voter must produce a ranked preference order of all candidates mentioned, and no ties are allowed. Such systems are called *ordinal*. However some voting systems, called *cardinal*, allow the voters to evaluate candidates separately, and a voter could say two candidates were equal. For the moment we shall concentrate on ordinal systems; cardinal systems will be studied in Chap. 7.

Given the ranked preferences of all voters, it is desirable to produce a list that represents the ranked preferences of the electorate. In other words, we would like to construct a list of all candidates in which, if A precedes B, then the electorate prefers A to B. (For convenience we write $A > B$ to mean A precedes B in the preference list.) We shall call a system that purports to produce such a list a *rank order voting system*. One example is Condorcet's extended method. Another is plurality voting: $A > B$ means A received more first-place votes than B, and the electorate ranks the candidates in order by the number of first-place votes received.

Not all ordinal voting methods are rank order voting systems; some simply find a winner and then the process stops. However, given any ordinal voting system, we can modify it so that it produces a ranking of any pair of candidates. For example, one could find the winner of an election, candidate A say. Then modify the preference profile by deleting A and recalculate the winner; say it is B. Delete B and continue. This will produce an ordered listing of all the candidates, which will be transitive.

But is it accurate?

In 1950, Kenneth J. Arrow [1] addressed this question, and proved that no rank order voting system for three or more candidates can convert the set of ranked preferences into a preference ranking for the whole electorate while meeting some very reasonable-sounding criteria. Arrow published his results in the book [2] in 1951, and a slightly improved version in the second edition in 1963 [3]. He subsequently received the Nobel Prize in Economics for his work in 1972.

© Springer International Publishing Switzerland 2014
W.D. Wallis, *The Mathematics of Elections and Voting*,
DOI 10.1007/978-3-319-09810-4_5

A quarter of a century after Arrow's first version was published, A. Gibbard [16] and M. A. Satterthwaite [25] independently proved a similar but stronger theorem that has become known as the *Gibbard-Satterthwaite Theorem*. Simply put, it essentially says that a dictatorship is the only voting system for three or more candidates that cannot be manipulated. We shall discuss their result in Sect. 5.5, below. For further discussion of strategy-proof voting, see [6].

5.1 Arrow's Criteria

Arrow originally stated five conditions that a fair system should satisfy. He subsequently modified them, resulting in three criteria: *No Dictators (ND)*, *Independence of Irrelevant Alternatives (IIA)* and *Pareto Efficiency(PE)*. We shall examine each of these.

No Dictators

This is the requirement that no single voter should have the power to determine the outcome. In a situation where one person (a "dictator") determines the result of an election, for example where the chairman of a company has the final say about all of its activities, voting would be a waste of time, so this condition is obvious. An election would only be held in those cases where the dictator declines to vote.

Say all voter rankings in a profile P remain fixed except for the ranking of one voter X, and however X relatively orders two candidates A and B, the electorate will order them in the same way. We shall say X is called *pivotal* for candidates A and B in the profile P. A voter is called *extremely pivotal* for candidate A in a profile if the voter is pivotal for all pairs involving A.

A voter who is pivotal for candidates A and B is also called a *pair dictator for A, B* in the profile P. Voter X is a *local dictator* for P if X is a pair dictator in P for all pairs. A *dictator* is a voter who is a local dictator for all profiles.

Independence of Irrelevant Alternatives

Say the election determines that the electorate as a whole prefers A to B, and suppose some electors change their preference lists. If no voter changes the relative positions of A and B—all those who initially ranked A ahead of B still do so in the final vote, and similarly those who preferred B to A continue to do so—then the system should continue to say that A is preferred to B.

Pareto Efficiency, or Unanimity

If every voter prefers A to B, then the system cannot say that the electorate prefers B to A.

Vilfredo Pareto (1848–1923) was an Italian engineer, sociologist, economist, political scientist and philosopher. An allocation of funds is called *Pareto efficient* if there is no other allocation in which some other individual is better off and no individual is worse off. The application of this term to voting systems is a little odd, but is now well-established.

Other Criteria

Arrow's original Theorem as stated in [1] involved two other criteria, rather than Pareto efficiency. These were *monotonicity* and *non-imposition*. Arrow changed them to Pareto efficiency in 1963, in [3].

Monotonicity (also called *mono-raise*, see [36]) states that if one or more voters change their ranked preferences by putting one candidate higher, then the overall preference list should either be changed by ranking that candidate higher or else be unchanged; an individual cannot be made *less* popular overall by having one rating *improved*. Non-imposition means that every possible overall preference list should be achievable: if there are n candidates, then each of the $n!$ lists can be achieved.

It is not hard to show that Independence of Irrelevant Alternatives, monotonicity and non-imposition together imply Pareto Efficiency, and in fact non-imposition is not required, but Independence of Irrelevant Alternatives and Pareto Efficiency do not imply monotonicity. So the set of conditions is weaker, and the Theorem is stronger, in the 1963 version, which is now referred to as Arrow's Impossibility Theorem:

Theorem 2. *No rank order voting system for three or more candidates that has no dictator can satisfy both Independence of Irrelevant Alternatives and Pareto Efficiency.*

The proof is in the next section.

Sample Problem 5.1 *Prove that Independence of Irrelevant Alternatives and monotonicity together imply Pareto Efficiency.*

Solution. Suppose a voting system is not Pareto efficient, but it satisfies Independence of Irrelevant Alternatives and monotonicity, Select a preference profile in which every voter prefers A to B, but the system says the electorate prefers B to A. Change the profile by moving B up in every voter's list, until B is just above A. According to monotonicity, this cannot lower B's ranking overall, so B will still be preferred to A. Now change every ranking by moving A down to the position originally occupied by B; by Independence of Irrelevant Alternatives, B is still preferred to A. But the rankings are now exactly as they were originally, with A and B exchanged, so the end result is to exchange the positions of A and B in the final result, and A is now preferred to B—a contradiction.

Practice Exercise. Prove that Pareto efficiency implies non-imposition.

5.2 The Proof of Arrow's Theorem

Several proofs are available. What follows is based on one of three given by Geanakoplos in [15].

We begin with the following result, called the *Extremal Lemma*.

Lemma 1. *Assume the voting system satisfies Pareto Efficiency and Independence of Irrelevant Alternatives alternatives, and suppose there is one candidate, C, such that every voter either places C at the top of the preference list or at the bottom. Then either C is elected, or else the electorate prefers every candidate to C.*

Proof. Suppose not. Say there is a profile in which every voter either places C at the top or at the bottom, but the system places C somewhere in the middle; say $A > C > B$. By the Independence of Irrelevant Alternatives, this would continue to be true even if every voter changed their preferences to place B above A—this will not disturb the relative positions of A and C, or of B and C in any ranking. Every elector prefers B to A, so by Pareto efficiency, the electorate prefers B to A. But the electorate holds $A > C > B$, so by transitivity A is preferred to B—a contradiction. $\square$

Proof of Arrow's Theorem. We assume that our voting system satisfies both Independence of Irrelevant Alternatives and Pareto Efficiency. We prove that it has a dictator.

Suppose there are n voters, $X_1, X_2, \ldots, X_n$. (The ordering is completely arbitrary.) Consider a profile in which every voter has placed candidate C at the bottom of the ranking. We shall call this Profile P_0. Construct a series of profiles, Profile P_1, Profile $P_2, \ldots$, Profile P_n, as follows. Profile P_1 is formed by changing X_1's preferences by moving C from the bottom to the top, leaving everything else unchanged. In general, Profile P_j is the same as profile Profile P_{j-1} except that X_j's preferences by moving C from the bottom to the top, with the rest of X_j's preference order unchanged.

From the Extremal Lemma we see that C is ranked either last or first in each of these profiles. By Pareto efficiency, C will be ranked last by the electorate under

P_1 and first—the winner—under P_n. Suppose P_k is the first profile in which C is the winner; C is ranked last under $P_1, P_2, \ldots, P_{k-1}$ and first under P_k. So X_j is extremely pivotal for C in profiles P_{k-1} and P_k.

We now show that X_j is a pair dictator in P_k for every pair not involving C. Suppose A and B are two other candidates. Select one of them, A say, and construct a new profile Q: X_j moves A above C, so that X_j ranks $A > C > B$, and let all other voters change their votes in any way provided that C stays in the same extremal position as it was in their vote in P_k. By Independence of Irrelevant Alternatives, the electorate will rank A above C (since all voters rank A and C in the same order as they did in profile P_{k-1}, where C was at the bottom overall), and it will rank C above B (since all voters rank A and C in the same order as they did in profile P_k, where C was at the top overall). So society ranks A above B whenever $X + j$ ranks A above B. But the same argument applies if B had been selected instead of A. So X_j is a pair dictator for every pair not involving C.

But the whole argument can be applied if some other candidate, D say, were used instead of C. (D could equal A, or B, or some other candidate.) Therefore there will be some voter, X_h say, who is a pair dictator for every pair not involving D. As $D \neq C$, X_h is a pair dictator for every pair involving C. In particular, we can assume D is neither A nor C. Then x_h is a pair dictator for A, C. If X_k and X_h are different, X_j cannot force a change in the electorate's relative ranking of A and C. But we have already seen that X_j can do just that, in the transformation from P_{j-1} to P_j. So X_k and X_h must be the same voter, and therefore be a dictator. So the voting system has a dictator.   □

5.3 Systems that Do Not satisfy IIA

It is easy to see that the Borda count does not always satisfy the Independence of Irrelevant Alternatives, even in a very small case. Consider three candidates and eight voters, with the profile

3	5
A	C
B	B
C	A

Using a 3-2-1 Borda count, C wins with 18 points; B receives 16 and A 14. However, if the three voters who preferred A decided that B was a better candidate, the profile would be

3	5
B	C
A	B
C	A

and B would be the winner, with 19 points compared to C's 18 and A's 11. No voter changed their preference ordering of B and C, but the electorate's ordering of those two candidates has changed.

The Hare system also does not satisfy the Independence of Irrelevant Alternatives. For example, consider

2	4	2	3
C	A	B	C
A	B	C	B
B	C	A	A

Under the Hare system, no one meets the quota of 6, so B is eliminated; then C wins 7–4. But if just two voters changed their preference list from $C > B > A$ to $B > C > A$, the profile becomes

2	4	4	1
C	A	B	C
A	B	C	B
B	C	A	A

and C is eliminated on the first count. Then A wins the election. No voter changed their ordering of A and C.

Bucklin's method does not satisfy the Independence of Irrelevant Alternatives either. To see this, consider the preference profile

5	5	2	3	5
A	B	C	C	D
B	C	B	D	C
D	A	A	A	B
C	D	D	B	A

The quota is 11. No one meets the quota initially, but after second preferences are added, C has 15 votes, B has 12, and the others do not meet the quota. So C is elected.

However, suppose the five voters with first preference B were to exchange A and C in their preferences. The profile is now

5	5	2	3	5
A	B	C	C	D
B	A	B	D	C
D	C	A	A	B
C	D	D	B	A

Again no one meets the quota initially, but after second preferences are added, B has 12 votes; no one else meets the quota, and B is elected. No voters changed the relative rankings of B and C.

5.4 Systems that Do Not Satisfy Monotonicity

It is not hard to see that the Hare system does not satisfy monotonicity. For example, consider the profile

1	4	6	5
A	A	B	C
B	C	C	A
C	B	A	B

No candidate has met the quota; A and C each have one vote, so both are eliminated, and B wins the election. However, if one of A's supporters were to promote B to first place, resulting in the profile

1	4	6	5
B	A	B	C
A	C	C	A
C	B	A	B

then A would be eliminated, and C would beat B by 9–7. Examples that do not involve a tie would involve more than one voter having changed their preference list; see Exercise 4.

Runoff voting does not satisfy monotonicity. For example, with the preference profile

11	2	7	4	4
A	B	B	C	C
B	A	C	A	B
C	C	A	B	A

the runoff would be between A and B, and A would win by 15–13; if the two voters represented by the second column were to change their ballots by moving A to first place, the resulting profile

13	7	4	4
A	B	C	C
B	C	A	B
C	A	B	A

would see B eliminated, and C would win the runoff 15–13.

5.5 The Gibbard-Satterthwaite Theorem

As we said earlier, A. Gibbard [16] and M. A. Satterthwaite [25] independently proved the *Gibbard-Satterthwaite Theorem*, Theorem 3 below, which essentially says that a dictatorship is the only voting system for three or more candidates that cannot be manipulated.

Gibbard-Satterthwaite uses the ideas of No Dictators and Unanimity, although unanimity is expressed slightly differently: a system is *unanimous* if, whenever candidate A is ranked first by every voter, then A will win. Instead of IIE, it uses the following definition: a voting system is *strategyproof* (or *non-manipulable*) if a voter can never improve the chances of her favorite candidate by strategic voting; a voter will always obtain the best result by ranking the candidates according to her true preferences.

For the rest of this section we shall assume that there are at least three candidates, and that the voting system satisfies the Pareto condition, and is unanimous and strategyproof. Each voter is required to have a preference list of all candidates, with no ties allowed. We shall prove several lemmas, leading to the Gibbard-Satterthwaite Theorem, Theorem 3. Our proof is based on the work of Taylor [29, 30]; see also [31]. Two other relatively simple proofs are given by Benoît [7] and Ninjbat [20].

Say S is some set of voters and A and B are two candidates. We say "S can use A to block B" if, whenever every member of S ranks A above B then B will not be elected. We denote this by $A >_S B$. If there is even one preference profile in which every voter in S ranks A higher than B but every voter not in S ranks B higher than A, and B is not elected, then S can use A to block B. We state this as a Lemma:

Lemma 2. *To show that S can use $A >_S B$, it suffices to produce a (single) preference profile in which (i) every voter in S ranks A higher than B; (ii) every voter not in S ranks B higher than A; (iii) B is not elected.*

Lemma 3. *Say A, B and C are three candidates, and say S is a set of voters such that $A >_S B$. Select a partition $S = P \cup Q$ of S into two disjoint sets. Then either $A >_P C$ or $C >_Q B$.*

Proof. We use Lemma 2. It suffices to produce a preference profile in which either

(i) Everyone in P ranks A higher than C, every other voter ranks C over A, and C is not elected, so from Lemma 2 $A >_P C$; or

(ii) Everyone in Q ranks C higher than B, every other voter ranks B over C, and B is not elected, so from Lemma 2 $C >_Q B$.

We consider a profile of the form

P	Q	R
A	C	B
B	A	C
C	B	A

where all voters place any candidates other than A, B, C lower than those three, all voters in a set have the same preference list, and R is the set of all voters not in S. Clearly the Pareto condition implies that A, B or C will be elected; moreover, every member of S prefers A to B, and $A >_S B$, so B will not be elected. Either A or C will win.

Say A is elected. Every voter in P ranks A over C and every other voter ranks C over A. So $A >_P C$. Conversely, if C is elected, every voter in Q ranks C over B and every other voter ranks B over C. So $C >_Q B$. $\quad\square$

Lemma 4. *Say A and B are two candidates, and say S is a set of voters such that $A >_S B$. Then S can use A to block any third candidate, and S can use that same third candidate to block B.*

Proof. In Lemma 3, suppose Q is empty, and $P = S$. Say the third candidate is C. Either $A >_P C$ or $C >_Q B$. But an empty set of voters cannot block a candidate. So P—that is, S—can use A to block C. If instead we take P to be empty, we see that S can use C to block B. $\quad\square$

Lemma 5. *If there are candidates A and B such that $A >_S B$, then $B >_S A$.*

Proof. Select a third candidate C. From Lemma 4, $A >_S C$. Now apply Lemma 4 again, but with the roles of B and C reversed; Since $A >_S C$, it follows that $B >_S C$. Finally, use Lemma 4 again; since $B >_S C$, S can use B to block any candidate, and in particular $B >_S A$. $\quad\square$

We shall call a set S of voters a *dictating set* if $A >_S B$ for every pair of candidates A, B. A one-element set of voters is a dictating set if and only if the member is a dictator. In any unanimous system, the set of all electors is a dictating set: all that is required is for all the voters to put A first, and A will block any other candidate.

Lemma 6. *If there are candidates A and B such that $A >_S B$, then S is a dictating set.*

Proof. Suppose C and D are any two candidates. We show that $C >_S D$. Since C and D are any two candidates, this means that S is a dictating set.

First, suppose $D = A$. We know from Lemma 5 that $B >_S A$, so Lemma 4 says that S can use any third candidate to block A. Taking C as the third candidate, we have the result. If $D \neq A$, Lemma 4 says that S can use A to block any third candidate, D say. So, using Lemma 4 again, S can use any third candidate to block D; this time, take C as the third candidate.   □

Lemma 7. *Say S is a dictating set of voters. Select a partition $S = P \cup Q$ of S into two disjoint sets. Then either P or Q is a dictating set.*

Proof. Say A, B and C are three candidates. Since S is a dictating set, $A >_S B$. From Lemma 3, either $A >_P C$ or $C >_Q B$. In the first case, Lemma 6 shows that P is a dictating set. In the second case, Lemma 6 shows that Q is a dictating set. □

Theorem 3. *Suppose there are at least three candidates, and the electoral system requires each voter to have a preference list of all candidates, with no ties allowed. If the system is unanimous and strategyproof, and satisfies the Pareto condition, then it is a dictatorship.*

Proof. Suppose there are n voters. Unanimity implies that the set of all voters is a dictating set. Partition the set of all voters into two non-empty sets, say P_1 and Q_1. It follows from Lemma 7 that either P_1 or Q_1 is a dictating set. Say it is P_1; this set will have at most $n - 1$ members. Applying Lemma 7 again, it follows that there is a proper subset of P_1 that is a dictating set, and that set will have at most $n - 2$ members. Continue in this way. After at most $n - 1$ iterations, we find that there is a dictating set with one element, that is, a dictator.   □

Exercises 5

1. Prove that plurality voting satisfies monotonicity.

2. Prove that plurality voting satisfies Pareto Efficiency.

3. Prove that the runoff method does not satisfy Independence of Irrelevant Alternatives.

4. Here is the profile for 15 voters choosing between three candidates A, B and C. The Hare system is to be used.

4	5	3	3
A	C	B	B
B	A	C	A
C	B	A	C

Use this profile to show that the Hare system does not satisfy the requirement of monotonicity.

5. Prove that the runoff method satisfies Pareto Efficiency.

6. Prove that the Hare system satisfies Pareto Efficiency.

7. Prove that the Condorcet system satisfies No Dictators, Independence of Irrelevant Alternatives and Pareto Efficiency. Why does this not contradict Arrow's Theorem?

8. Use the preference profile

5	8	7
A	B	C
B	C	A
C	A	B

to prove that the Coombs rule does not satisfy Independence of Irrelevant Alternatives.

9. Use the preference profile

2	3	5
A	B	C
B	C	B
C	A	A

to prove that Bucklin's method does not satisfy Independence of Irrelevant Alternatives.

10. Does pairwise sequential voting satisfy Independence of Irrelevant Alternatives?

11. Recall that an electoral system satisfies *non-imposition* if, for every candidate, there exists a preference profile for which that candidate is the winner. Prove that if a system is Pareto efficient then it must also satisfy non-imposition, but that the converse is not true.

12. Prove that the Gibbard-Satterthwaite Theorem still holds if we replace the Pareto criterion by non-imposition.

Chapter 6
Complex Elections

In this Chapter we discuss cases where several candidates are to be elected simultaneously. Such elections are called *complex* or *multiple* elections.

6.1 The Generalized Hare Method

The Hare system was modified by Andrew Inglis Clark, who was Attorney-General of Tasmania in the late nineteenth century. This *generalized Hare method*, or *quota method*, is used in many countries, including Ireland, Australia and Malta.

The method works as follows. Each voter makes a *preferential* vote, providing a complete list of all the candidates, in order of preference (a *preference list*). The first name on the list is called the voter's *first preference*, and so on. All first preferences are counted; if candidate A receives at least a certain number of votes (called the *quota*) then A is elected, and the unneeded votes, called the *excess*, are distributed proportionally among the second preferences of the electors who chose A. Then the new set of votes is examined again, to see if any more candidates have no achieved the quota after the additional votes are added. If at any stage no one new has been elected, the candidate with the fewest first-place is eliminated, and all those votes go to the second preferences. If, at any stage, the number of candidates not yet elected or eliminated is equal to the number of seats remaining, all those candidates are elected.

As an example, suppose 24,000 people are to elect five representatives from a larger number of candidates, and quota is set at 4,001 votes. (This number is chosen because you cannot have six candidates who each get more than 4,000 first preferences.) Every candidate who gets more than 4,000 votes is elected. If a candidate

© Springer International Publishing Switzerland 2014
W.D. Wallis, *The Mathematics of Elections and Voting*,
DOI 10.1007/978-3-319-09810-4_6

gets more than 4,000 votes, the excess go to voters' second choices, divided proportionally. To illustrate this, suppose A gets 5,000 votes. Of these, 2,500 have B as second choice, 2,000 have C and 500 have D. Since 5,000 is greater than 4,000, A is declared elected.

The 1,000 surplus votes are divided in the proportion 2,500:2,000:500, or 50 % : 40 % : 10 %. That is, B gets 50 % of A's excess, because 2,500 is 50 % of 5,000; C gets 40 %; and D gets 10 %. As A has 1,000 extra votes, we say she has 1,000 preferences to be distributed. B gets 50% of these, so 500 more votes are added to B's total. In the same way C gets 400 added votes (40 %) and D gets 100 votes (10 %).

Now we check whether B, C or D has met the quota. For example, if B previously had 3,700 votes, the new total would be 4,200, exceeding the quota again, so B is declared elected.

In general, suppose there are V voters and N places to be filled. We shall define the *lower bound* for this election to be $\frac{V}{N+1}$, and the quota is the smallest integer greater than the lower bound. A candidate is declared to be elected if his or her number of votes exceeds the lower bound, and the excess is found by subtracting the quota from the candidate's total number of votes. Notice that, if only one candidate were to be elected, the quota requirement is that a candidate receive a majority of the votes.

This quota is called the *Droop quota*, and was suggested by H. R. Droop [13]. The original quota, called the *Hare quota*, is simply $\frac{V}{N}$. (The excess under the Hare quota is found by subtracting the quota from the number of votes) An interesting comparison of the two quota systems can be found in [32].

Some systems use the formula $\frac{V}{N+1} + 1$ for the quota, and say a candidate is elected if the quota is equalled or exceeded. This may require one more vote than the Droop quota in the case where $\frac{V}{N+1}$ is not an integer: not important in most political elections, but it may be significant if a small group is voting.

In a real example, the whole list of preferences is kept. The process may require a great deal of data; moreover it may result in complicated numbers, fractions, and so on. Historically this was a serious problem and caused long delays in announcing the results of elections, but it is no longer an issue now that computers are available and voting machines can be adapted to keep all the data.

For example, given a preference profile of the form

12	9	6	...	...
A	A	A	B	...
B	B	C	A	...
C	D	B	D	...
D	C	D	E	...

and a quota of 19, A's surplus is nine. As B received 21 of the second preferences of A's voters and C received 6, the surplus is divided in the ratio 21 : 6 between B and C; B receives 7 further votes and C receives 2. Another way of looking at this is to say the votes for A were distributed 12 : 9 : 6 among the three preference groups, so the 9 surplus votes should be distributed in the same ratio, which comes to 4, 3 and 2 votes respectively. This will make later calculations easier. To look at the second candidate to be elected, delete A from every preference list and replace the votes in those columns where A was originally the first choice with the surplus amounts, so that the profile becomes

4	3	2	. . .	. . .
B	B	C	B	. . .
C	D	B	D	. . .
D	C	D	E	. . .

We carefully chose the numbers of votes in the above example so that the surpluses allocated were all integers. This was done for simplicity, but is not essential. Suppose the numbers of votes in the first three columns had been 14, 8 and 5, which still gives a total of 27 and a surplus of 9. In order to find the second candidate elected, the votes allocated to the first three columns would be $\frac{14}{27} \times 9$, $\frac{8}{27} \times 9$ and $\frac{5}{27} \times 9$, or $4\frac{2}{3}$, $2\frac{2}{3}$ and $1\frac{2}{3}$. The arithmetic may become more complicated, but in the real world the calculations will be carried out by computer, so there is no problem.

Sample Problem 6.1 *Say there are five candidates for three positions; the preference table is*

6	6	9	6	3	2
A	A	C	C	E	E
B	B	D	D	C	A
E	D	E	E	D	B
D	E	A	B	A	C
C	C	B	A	B	D

Who will be elected?

Solution. There are 32 voters, so the lower bound is 8 and the quota is 9. A gets 12 first place votes and C gets 15, both of which meet the quota. So A and C are elected.
A has a surplus of 4. The second-place candidate in every case is B, but we observe that the votes are divided in proportion 6:6 between A's two lists, so we give two votes to each of A's lists. C is also at the top of two lists, and has a surplus of 7, divided 9:6. This gives surplus allocations of 4.2 and 2.8. After these allocations, the new table is

2	2	4.2	2.8	3	2
B	B	D	D	E	E
E	D	E	E	D	B
D	E	B	B	B	D

B has four votes, D has seven, and E has five. No one meets the quota. (Note that the quota does not change.) So B (who had the fewest votes) is eliminated. We now have

2	2	4.2	2.8	3	2
E	D	D	D	E	E
D	E	E	E	D	D

D now has 9 votes and E has 7, so D is elected. In total, A, C and D are elected.

Practice Exercise. Repeat the above problem for preference profile

5	5	7	8	5	2
A	A	B	C	D	E
B	E	D	D	C	A
E	B	E	E	E	B
D	D	A	B	A	C
C	C	C	A	B	D

The choice of quota can make a significant difference. As an example, consider an election in which five candidates are to be elected, and two political parties are involved. (This was the situation in state elections for the federal senate in Australia, in the middle of the twentieth century.) Typically at least two candidates are elected from each of the two major political parties, with the fifth member being from one of the two parties, from a third party, or an independent. For simplicity we shall assume there are only six candidates, A, B and C from one party (the Republicrats) and D, E and F from the other (the Democans), no third party or independents. We shall assume there are 240 voters, and all electors belong to one or other party. The Hare quota is 48 and the Droop quota is 41. The preference profile is

65	55	4	42	42	32
A	B	C	D	E	F
B	A	A	E	D	D
C	C	B	F	F	E
D	D	D	A	A	A
E	E	E	B	B	B
F	F	F	C	C	C

As you can see, the Republicrat party has 124 votes and the Democans 116, so we would expect all three Republicrats and two Democans to be elected, and this happens when the Droop quota is used: A, B, D and E all achieve the quota, and after the excesses are distributed C has 44 votes to F's 36. However, using the Hare quota, we see that A and B are elected and the new profile is

17	7	4	42	42	32
C	C	C	D	E	F
D	D	D	E	D	D
E	E	E	F	F	E
F	F	F	C	C	C

No candidate is elected, so C, with only 28 votes, is eliminated, and D, E and F are elected.

6.2 The Generalized Coombs Rule

The Coombs rule can also be applied to cases where more than one candidate is to be elected. Again, each voter provides a complete preference list. The Droop quota is used. Again, any candidate who receives at least as many votes as the quota is declared elected and removed from the preference profile, and the excess votes are distributed proportionally.

If no candidate has achieved the quota, the candidate with the greatest number of *last place* votes is deleted, and the votes are counted again. If there are two such candidates (with equal numbers of last places), both are eliminated.

This method often provides the same result as the Hare method, but not always. Here is an example with four candidates, where two are elected under one system but the other two are the winners under the other system.

Sample Problem 6.2 *Say there are four candidates for two positions; there are 126 voters, and the preference table is*

36	35	37	18
A	B	C	D
B	C	D	A
C	D	A	C
D	A	B	B

Who will be elected under the Hare system? Who will be elected under the Coombs rule?

Solution. The quota is 43. No candidate gets enough first-place votes to meet the quota.

Under the Hare method, D, with only 18 votes, is eliminated. The new table is

36	35	37	18
A	B	C	A
B	C	A	C
C	A	B	B

A has 54 votes, and is declared elected, with a surplus of 12. Eight of these twelve votes are allocated to the first column and four to the fourth. The result is

8	35	37	4		43	41
B	B	C	C	=	B	C
C	C	B	B		C	B

and *B* is elected. So *A* and *B* are the Hare winners.

Under the Coombs rule, *B* is the first candidate to be eliminated, with 55 last-place votes. The new table is

36	35	37	18		36	72	18
A	C	C	D	=	A	C	D
C	D	D	A		C	D	A
D	A	A	C		D	A	C

and *C*, with 72 votes, has achieved the quota. The surplus is 30, and it all goes to the second column, so the new table is

36	30	18		36	48
A	D	D	=	A	D
D	A	A		D	A

and *D* is the second candidate elected.

Practice Exercise. Who are the winners of the election presented in Practice Exercise 6.1, if the Coombs rule is applied?

6.3 The Single Transferable Vote

The *single transferable vote* (or STV) system is a variation of the generalized Hare method that contains some aspects of approval voting, a system that we shall examine in the next chapter. Every elector decides whether or not a candidate is approved, and supplies a preference profile of the approved candidates. In the preference table, some columns will have blank spaces at the bottom. In some countries the name "single transferable vote" is also applied to the Hare method and generalized Hare method, but we shall restrict it to the case where a voter is allowed to omit a candidate.

There are various ways in which an STV system can be implemented. Typically, any vote that lists *no* approved candidates is eliminated before the quota is calculated. Then elections proceed similarly to elections under the generalized Hare method. However, it is possible for a vote to disappear because all the candidates named on it have been eliminated. In the basic system this can mean that no candidate reaches the quota, even though not all vacancies have been filled; the standard procedure is to hold another election for the remaining seats. However, another possibility is to recalculate the quotas at each stage. We shall refer to this as *dynamic* STV.

Sample Problem 6.3 *An election for two positions is held under the single transferable vote system; the preference table is*

5	5	10	3	1
A	A	A	B	C
B	C		C	B
C	B		A	A

What is the result under the basic STV system? What is the result under dynamic STV?

Solution. There are 24 electors, so the quota is 9. A is elected and has 12 surplus votes. The new preference table is

3	3	6	3	1
B	C		B	C
C	B		C	B

, or equivalently

6	4	6
B	C	
C	B	

B has six votes and C has four; neither is elected and a new election is held for the remaining position.

Under the dynamic system, the new preference table is treated as if it were the preference table for a new election; the "empty" ballots are discarded, and we proceed as though there were ten votes with preference table

6	4
B	C
C	B

B is elected.

Practice Exercise. In the above problem, suppose the voters all decided to vote for at least two candidates, and the preference table was

5	5	5	5	3	1
A	A	A	A	B	C
B	C	B	C	C	B
C	B			A	A

What is the result under the basic STV system? What is the result under dynamic STV?

Exercises 6

*In Exercises **1– 4**, a preference table is shown for an election where two candidates are to be elected using the generalized Hare method. In each case, what is the quota? What is the outcome of the election? What would be the outcome using the Coombs rule?*

1.

7	8	7	5
A	B	C	D
B	D	B	C
C	C	A	B
D	A	D	A

2.

4	6	6	6	2	6
A	D	B	E	C	B
D	A	D	A	E	A
E	E	C	D	D	E
B	B	E	C	A	C
C	C	A	B	B	D

3.

3	5	8	8	6
A	A	B	C	D
B	C	D	B	C
C	D	C	A	B
D	B	A	D	A

4.

6	7	7	7	7	5
A	B	C	D	C	E
E	D	B	E	D	B
D	A	D	A	E	A
B	E	E	C	A	C
C	C	A	B	B	D

5. In Exercises **1** and **3**, what would be the quota if the Hare quota were used instead of the Droop quota? Would the result be changed?

6. In Exercises **2** and **4**, what would be the quota if the Hare quota were used instead of the Droop quota? Would the result be changed?

*In Exercises **7– 12**, a preference table is shown for an election where three candidates are to be elected using the generalized Hare method. In each case, what is the quota? What is the outcome of the election? What would be the outcome using the Coombs rule?*

7.

4	7	9	8	7	5
A	A	B	C	D	D
B	B	D	D	C	A
C	D	C	A	B	B
D	C	A	B	A	C

8.

8	5	6	7	4	3
A	B	C	D	E	E
B	A	D	C	C	A
E	D	E	E	D	B
D	E	B	B	B	D
C	C	A	A	A	C

9.

6	6	6	6	5	3
A	A	C	C	E	E
B	D	D	E	C	A
E	E	E	D	D	B
D	B	A	B	A	C
C	C	B	A	B	D

10.

9	8	7	6	4	2
A	B	D	D	C	A
E	D	E	E	D	B
D	A	C	A	E	E
B	E	B	C	A	C
C	C	A	B	B	D

11.

6	6	7	5	6	5	5
A	B	B	C	D	E	E
B	E	D	E	E	B	B
D	A	A	B	A	C	A
C	C	E	A	B	D	D
E	D	C	D	C	A	C

12.

8	6	7	7	9	3	8
A	A	B	C	C	D	E
D	B	C	D	E	C	A
E	D	A	E	D	B	B
B	E	D	A	A	E	D
C	C	E	B	B	A	C

13. Suppose a generalized Hare election were held using a quota smaller than the Droop quota. Construct an example to show that there could be no result.

14. An election is to be held under the STV system, and three candidates are to be elected. The preference table is

7	7	4	2	4	6
A	B	C	D	B	F
B	A	B	B	A	C
		D	A	E	E
		A	C	D	
		E	F		

(i) What is the quota?

(ii) What is the result under the basic STV system?

(iii) What is the result under dynamic STV?

15. In an STV election, 39 voters must elect two candidates and three candidates are to be elected. The preference table is

6	6	4	4	7	3	6	0
A	A	B	B	C	C	D	
B		A	D	D	B	A	
C		C	A	A	D	B	
		D				C	

(i) What is the quota?

(ii) What is the result under the basic STV system?

(iii) What is the result under dynamic STV?

Chapter 7
Cardinal Systems

As we pointed out in Chap. 4, cardinal systems allow individual votes to include ties. They also allow the electorate to express overall approval or overall disapproval of a candidate. For that reason, they are often used in situations where the number of candidates to be elected is not determined in advance, and sometimes where it is possible for no one to be elected.

7.1 Approval Voting

Approval voting was first used in cases where the number of people to be elected was not fixed. For example, if a committee wants to co-opt a few more members to help organize a function, they might add all those who they think would make a positive contribution, not a fixed number. In that case a member might vote for every suitable candidate, casting as many votes as she wishes; all the votes are considered equal. The number of votes received by a candidate is called his *approval rating*. Candidates with an approval rating of at least 50 % (or 60 %, or some other agreed figure) are elected.

Sometimes there is a strict requirement that a certain number be elected, or a minimum or maximum is imposed. Ties are possible, and some sort of runoff procedure may be necessary. The following example illustrates these ideas.

Sample Problem 7.1 *Ten board members vote on eight candidates by approval. The candidates are A, B, C, D, E, F, G, H, and the board members are $q, r, s, t, u, v, w, x, y, z$. They vote as follows ($\times$ represents approval).*

© Springer International Publishing Switzerland 2014
W.D. Wallis, *The Mathematics of Elections and Voting*,
DOI 10.1007/978-3-319-09810-4_7

	q	r	s	t	u	v	w	x	y	z
A	×	×	×		×		×	×		×
B		×	×	×	×	×	×	×	×	
C			×					×		
D	×		×	×	×	×	×	×		×
E	×	×	×	×	×			×		
F	×	×	×	×	×	×	×		×	×
G	×		×	×			×		×	
H		×		×	×	×				×

(i) Which candidate is chosen if just one is to be elected?

(ii) Which candidate is chosen if the top four are to be elected?

(iii) Which candidate is chosen if the top two are to be elected?

(iv) Which candidate is chosen if at most four are to be elected and 80 % approval is required?

Solution. Here is a summary of the votes received:

$$A - 7 \quad B - 8 \quad C - 2 \quad D - 8 \quad E - 7 \quad F - 9 \quad G - 5 \quad H - 7.$$

(i) F, who received the most votes.

(ii) B, D, F are elected; there must be a runoff election between A, E and H.

(iii) F is elected; there will need to be a runoff election between B and D.

(iv) $B \ D$ and F are elected.

The array in the example is called an *approval table*; we shall always denote approval with a cross in such tables.

Approval voting is particularly useful for situations like the selection of new employees. In those circumstances there is usually a minimum requirement, and further applications may be called if not enough good applicants are available.

On the other hand, there are arguments both for and against the use of approval voting in national politics. For a full discussion (basically in favor of approval voting), see [10]. One argument is that approval voting would add support to third party candidates, which could eventually undermine the two-party system in countries such as the United States; whether this would be a good thing or a bad thing is a matter of opinion.

One place where approval voting has been used is in primary elections. One example, also in [10], is the 1980 Republican primaries in New Hampshire. In that election, the results were Reagan 50 %, Bush 23 %, Baker 13 %. Baker dropped out of the primaries 1 week later, after two more poor showings, and eventually the party endorsed Reagan for president and Bush for vice-president. However, a poll was taken among the New Hampshire voters to determine how they would have cast approval votes. Reagan still won, with 58 %, but Baker gained 41 % to Bush's 39 %. So approval voting might have resulted in a different vice-presidential endorsement, which would certainly have affected the 1988 presidential race.

7.2 Range Voting

In approval voting, the electors do not distinguish between the candidates of whom they approve, but in the real world we would usually prefer one candidate over another, even though we found both of them acceptable. A type of approval voting that incorporates this idea is *range voting*, also called *score voting*. In the simplest form, all voters allocate a score to each candidate, the scores are added, and the candidate with the highest score wins. The most common examples are sports and contests with a number of judges; recent examples have included a number of television games. Essentially the same technique is used for ranking movies on IMDB and ranking books, hotels and restaurants on internet sites.

More complicated versions exist. In springboard and platform diving, three or more judges award to each dive a score from 1 to 10, with half points allowed. The scores are added for the dive and then multiplied by a constant, the *degree of difficulty*, that depends on the type of dive. Finally all scores are added for each diver.

When there are five diving judges, the usual method is to delete the highest and lowest scores before adding. This is called *truncation*. In most international competitions there are seven or nine judges; the dive score is calculated by adding the middle five awards, multiplying the sum by the degree of difficulty, and then multiplying by 0.06. The result is the equivalent of a three-judge score.

A form of range voting was used for elections in ancient Sparta [14]. The crowd would cheer for the candidates, and the one who received the loudest reception would be elected, This technique is used in some sporting events and television shows.

Sample Problem 7.2 *Townspeople are voting on the site for a new football field. Sites are available in the North, West, East and South suburbs. The northern site is inconvenient for all citizens who live in other areas. All voters prefer the site nearest to them. Assume they allocate points out of 10, and (by a strange coincidence) all voters in an area allocate the points the same way, as follows (the percentage of the population in each area is shown):*

	North 40 %	West 24 %	East 16 %	South 10 %
North	10	0	0	0
West	4	10	6	5
East	2	4	10	7
South	0	2	6	10

Which area would be chosen:

(i) by majority;   *(ii) by plurality;*   *(iii) by Hare;*

(iv) by runoff;   *(v) by range voting?*

Solution. (i) No decision; (ii) North; (iii) South is eliminated, their votes go to East; West is eliminated; East wins; (iv) the runoff is between North and West, and West wins; (v) calculating the scores as though there were 100 voters, North scores 400, West 596, East 476, South 344, and West wins.

It is easy to see that range voting satisfies monotonicity: raising your score for a candidate cannot hurt their chance of winning, and lowering it cannot make the chance greater. It also satisfies Independence of Irrelevant Alternatives. There are groups that propose range voting for political elections.

However, range voting has two problems. First, it requires voters to make a serious assessment of the relative quality of candidates, rather than just deciding which is better; and second, even if the voters do this, it is not clear that "7 out of 10" will mean the same thing to different voters. In diving, for example, judges are required to undergo training. The idea of requiring all voters for the House of Representatives to undergo training to bring uniformity to their votes is not only repugnant to many people, but would be very expensive.

Moreover, range voting is open to manipulation; voters may give lower approval ratings to candidates that they consider second-best, in order to boost their favorite's chances. See, for example, Exercise 9.

The supporters of range voting have advocated it very strongly (see, for example, [23]). However, according to Wikipedia [35], no elected official in the United States is known to endorse range voting.

7.3 Cumulative Voting

A variant of range voting is called *cumulative voting*. It is used in over 50 communities in the United States, typically to elect bodies such as city councils and school boards (although it was used to elect the Illinois House of Representatives from 1870 to 1980). It is also used by corporations, and by committees voting to decide on various courses of action. In most cases, the method is used when several candidates are to be elected. We shall look at three types of cumulative voting.

In *equal and even cumulative voting*, each elector has one vote, but may split that vote among various candidates. The vote is divided evenly: for example, if a vote is split between four candidates, each candidate receives 0.25 of a vote. The votes are totalled for each candidate, and the candidates with the highest scores are elected.

In a typical community election, each elector is allocated the same number of votes, typically equal to the number of seats to be filled. You can allocate your votes as you wish; for example, if three people are to be elected, each elector receives three votes, and can allocate all three to one candidate, or two to one candidate and one to another, or one each to three candidates. The tactic of allotting many of your votes to one candidate is called *plumping*.

Another common use of cumulative voting is commonly used in corporations, for example to elect directors. Each voter is given a number of votes, also called *points*, and the number varies from voter to voter. Often the voters are shareholders, and the number of points equals the number of shares held. The voter can allocate points to different candidates in whatever proportion they wish. Then the election is conducted essentially using the generalized Hare method. This type of cumulative voting is called *points voting*.

Often a group of shareholders of a company will pool their votes, trying to form a faction large enough to elect a majority; they may also try to influence other shareholders. Suppose there are V votes to be cast and D directors are to be elected. The quota for electing a director is $\frac{V}{(D+1)} + 1$ votes. If a faction is to elect its N favorite candidates, it will need at least $\frac{VN}{(D+1)} + 1$ shares, because $\frac{V}{(D+1)}$ votes are deleted from the total each time a candidate is elected. Of course, some other voters may also vote for some of the faction's favorites, so a smaller number of votes may suffice; but $\frac{VN}{(D+1)} + 1$ shares guarantee the election of the selected N candidates. From the formula, it follows that possession of X shares can guarantee the election of $\lfloor \frac{(D+1)(X-1)}{v} \rfloor$ directors.

Cumulative voting is discussed at length in [33]. That web page includes further discussion of tactical voting in a cumulative system.

Exercises 7

1. Eight administrators s, t, u, v, w, x, y, z are voting to fill one position in senior management from candidates A, B, C, D, E, F. They use an approval system; their approval table is

	s	t	u	v	w	x	y	z
A	×	×				×		
B	×		×	×	×			×
C		×		×			×	×
D		×		×	×		×	
E			×				×	×
F		×	×	×			×	×

(i) What is the outcome?

(ii) What is the outcome if at least 80 % approval rating is required?

(iii) What is the outcome if at least 60 % approval rating is required?

(iv) The administrators are told that they can appoint more than one person if they wish. What is the outcome if at least 60 % approval rating is required?

In Exercises 2–7, an approval table is shown for selection of at most three candidates. In each case:

(i) What is the outcome if there is no minimum requirement?

(ii) How many votes are needed if at least 66 % approval is required?

(iii) What is the outcome if at least 66 % approval is required?

2.

	r	s	t	u	v	w	x	y	z
A	×	×				×			
B	×		×	×	×	×			×
C		×		×			×	×	×
D		×		×	×	×		×	
E			×	×			×	×	
F	×		×	×	×			×	×
G			×	×			×	×	
H	×		×	×	×			×	×

3.

	p	q	r	s	t	u	v	w	x	y	z
A	×	×		×	×	×		×			
B	×	×	×		×	×	×	×			×
C		×		×		×			×	×	×
D		×		×	×	×		×		×	
E	×	×	×	×	×	×	×		×	×	×
F		×	×					×	×	×	
G		×			×	×			×	×	
H	×		×		×	×	×			×	×

4.

	p	q	r	s	t	u	v	w	x	y	z
A		×	×	×	×	×	×	×		×	
B		×	×		×	×		×			×
C									×	×	×
D		×		×	×	×		×		×	
E	×		×	×					×		×
F				×				×	×		
G						×	×		×		
H	×	×	×	×	×	×	×			×	×

5.

	q	r	s	t	u	v	w	x	y	z
A	×		×		×		×	×		
B	×			×	×	×	×			×
C		×	×		×	×		×	×	×
D		×		×	×	×	×		×	
E	×	×	×	×	×	×		×	×	×
F		×	×				×	×	×	
G		×			×	×		×	×	
H	×		×				×		×	×
I		×			×	×		×		
J	×			×	×	×	×			×

6.

	q	r	s	t	u	v	w	x	y	z
A	×			×	×	×	×			×
B		×		×	×		×	×	×	
C			×		×	×	×		×	
D	×	×	×		×		×	×		
E	×		×	×	×	×		×		×
F			×				×		×	×
G			×	×	×			×	×	
H	×		×		×	×		×		
I	×	×	×		×	×		×		×
J				×	×	×	×			×

7.

	o	p	q	r	s	t	u	v	w	x	y	z
A		×	×			×	×	×	×			×
B		×		×		×	×		×	×	×	
C			×		×	×	×		×		×	
D	×	×		×	×		×		×	×		
E	×		×		×	×		×		×		×
F			×						×		×	×
G	×	×			×		×			×	×	
H	×		×		×	×					×	
I	×		×	×	×		×	×		×		×
J			×			×		×	×			×
K				×	×	×	×	×	×			×
L		×	×	×	×	×	×		×	×	×	

8. Suppose three candidates are to be elected. Voters have the following preference profile; if a candidate appears on a list, one may assume the voter approves of that candidate.

3	5	7	4	8	5
A	B	C	D	G	E
E	D	F	F	D	B
D	F	D	A	E	A
B	E	E	C	F	
		A	B	B	
			G	C	
			E	A	

(i) What is the result under approval voting?

(ii) What is the result under the basic STV system?

(iii) What is the result under dynamic STV?

9. Four voters V_1, V_2, V_3 and V_4, are to vote for one of alternatives A, B and C, using range voting with a maximum of five per candidate. Their approval levels are:

	V_1	V_2	V_1	V_4
A	5	2	1	0
B	4	4	1	5
C	1	5	3	4

Who will be elected? How could one voter who prefers C ensure that his candidate is elected?

10. A range election is held between representatives of the Democrats, Greens and Republicans. 103 voters express their preferences as follows, where the weights must lie between 1 and 5. The numbers of voters with a given preference list is shown.

Number	36	8	30	9	7	13
D	5	5	2	4	4	2
G	2	1	4	2	5	5
R	1	3	5	5	1	3

(i) Who would win under range voting?

(ii) Who would win under the Hare system?

(iii) Is there a Condorcet winner?

11. Thirty-five voters elect two representatives from a field of four. The voters express their preference as shown, with weights between 1 and 10, and a range election is conducted. The numbers of voters with a given preference list is at the top of each column. Who is elected?

Number	12	8	9	6
A	10	2	1	3
B	4	3	10	1
C	8	10	7	2
D	3	1	8	10

(i) Who is elected under range voting?

(ii) Who would win if the generalized Hare method is used?

(iii) It is decided to use approval voting; a candidate is approved if the range score is greater than 4, and the two best candidates are accepted. What is the result?

(iv) What would be the result of approval voting if approval was defined as "at least 5" and 50 % approval was required?

12. A company uses the cumulative voting to elect its the members of its board of directors. The company has 10,000 shares. If a given shareholder wishes to assure control of three directors out of nine up for election, what is the minimum number of shares that she must hold?

13. A company has ten directors and is holding an election for all ten positions. Assuming there are 18,000 shares, how many shares would a faction need to ensure election of a majority of directors?

Chapter 8
Weighted Voting

Most elections involve electorates sufficiently large that ties are extremely unlikely. However, some situations, particularly committees, involve small numbers of voters. In this chapter we shall discuss some of these cases.

8.1 Power

The concept of power in voting systems is well-known. For example, the United States Senate has 100 members. If the senators are voting on a matter where the political parties have no particular interest, we might say that each member has 1 % of the power of the body. On the other hand, many issues are viewed differently by different parties; if the Senate consisted of 62 Democrats, 37 Republicans and one Independent, we might say the Democratic Party had 62 % of the power; if the issue was such that all members of a party would definitely vote the same way, the Democrats would have 100 % of the power.

8.2 The Shapley-Shubik Power Index

In 1954, Lloyd Shapley and Martin Shubik [26] discussed voting power of individuals in committees, and invented a measure called the *Shapley-Shubik power index*.

Consider a situation where n voters are to vote on an issue, independently of any party commitment. We could rank the voters in terms of their commitment to the issue, with the voter most in favor of it first and the one most opposed to the

© Springer International Publishing Switzerland 2014
W.D. Wallis, *The Mathematics of Elections and Voting*,
DOI 10.1007/978-3-319-09810-4_8

issue last. After some discussion, the committee votes on a motion to endorse the issue. Let us assume that the relative ranks of the voters are unchanged; the only question is, how many of voters would vote in favor of the issue?

As an example, suppose there are five voters A, B, C, D, E, where A is most in favor, B second, and so on. (Call this the *preference order*.) Assume a majority of votes—that is, three votes—is required to pass the motion. In order to know the result, we need to know how C will vote on the issue. We say C is *pivotal*. In many cases we would know whether the motion would pass without consulting C. For example, if we know that D is in favor, we are sure that it will pass. The only case in which C's vote is important is when A and B are in favor and D and E are opposed. In that case, we say C has power over the motion.

There are 120 possible orderings for a 5-vote motion. Assuming a majority is required, the third voter in order is the one with power. So each voter would have power in 24 cases, or $\frac{1}{5}$ of the orderings. In this case we would say that each voter's Shapley-Shubik power index is $\frac{1}{5}$, or 20 %.

There are many more complicated power situations. For example, suppose the committee has four members. When there are two people in favor and two against, the Chair of the committee decides the motion; this is called a *casting vote*. Say the Chair is X and the other committee members are A, B, C. If the Chair is second or third on the preference order, for example $AXBC$ or $BCXA$, then the vote will go the way s/he votes; the Chair will have power in those orderings. So the Chair has power index $\frac{12}{24}$, that is $\frac{1}{2}$. The other three members have equal power index, $\frac{1}{6}$.

8.3 The Power of the President

Suppose both chambers of the United State Congress pass a piece of legislation; a majority must approve it in each chamber. If the President approves the bill, it becomes law. If not—if the President vetoes the bill—it will become law only if each house passes it by a two-thirds majority. So how much power does the President have?

There are 435 members of the House of Representatives, so a majority is 218 and two-thirds of the members is 290. For the Senate, with 100 members, the numbers are 51 and 67. Suppose the 536 voters—the House and Senate members and the President—are ordered according to their opinion on the bill. In order to be pivotal, the President would need to be preceded in the ordering by at least half the members of each chamber but by fewer than two-thirds of the members of at least one chamber.

Suppose the President is preceded in the ordering by x House members and y senators. The members who precede the President can be arranged in $(x+y)!$ ways, and those following in the ordering in $(535 - x - y)!$ ways. Moreover, the House members and the senators who precede the president can be chosen in $\binom{435}{x}$ and $\binom{100}{y}$ ways respectively. Therefore there are

$$N_{x,y} = (x+y)! \times (535 - x - y)! \times \binom{435}{x} \times \binom{100}{y}.$$

such orderings. The total number of orderings in which the ordering is pivotal is found by summing $N_{x,y}$ over all cases where $x \geq 218$ and $y \geq 50$, but it is not true that both $x \geq 290$ and $y \geq 67$. So the total number of orderings in which the President is pivotal is

$$\sum_{x=218}^{435} \sum_{y=50}^{100} N_{x,y} - \sum_{x=290}^{435} \sum_{y=67}^{100} N_{x,y}.$$

There are 536! orderings to the voters in total, so the fraction of the orderings in which the President is pivotal is found by dividing the above sum by 536!; using online binomial coefficient and factorial calculators (such as [21] and [24]) or a computer program (such as *Mathematica*), the answer is found to be approximately 0.16. So the President's Shapley-Shubik power index is about 16 %.

An interesting webpage on Presidential vetoes is [17].

8.4 Voting Blocs

In this section we shall look at one particular case, where one voter has a number of votes and all others have exactly one vote. The most common example of this is when one person has control of a number of others' votes; this is called a *proxy*. The set of voters will always vote the same way on any issue is called a *bloc*. In a later section we shall look at cases where the set of voters is not so rigidly fixed; this is called a *coalition* and is more often seen in political examples.

Voting blocs can be very powerful. For example, consider a committee of five members who vote using majority rule; the quota in every case will be three votes, and every member has power index $\frac{1}{5}$. Now suppose two of the members form a bloc, and one of them authorizes the other member to make all their votes. This is mathematically the same as the four-member committee with a chairman that we discussed in the preceding section. In that case, the Chair had half the power, and the other three had power index $\frac{1}{6}$ each. In the present case, we would divide the Chair's power between the two members of the bloc, so they each get $\frac{1}{4}$. In other words, they each have 25 % of the power, while the others have just under 17 %. But the difference can be much greater, as the next example shows.

Sample Problem 8.1 *A committee of seven members requires a simple majority of four votes to pass a motion. What are the power indices if there is a bloc with three members? What if there is a bloc of four members?*

Solution. Assume there is a three-member bloc X. X is treated as if it were a single voter with three votes, so there are $5! = 120$ ways to rank the voters. In the 24 orderings of type $XABCD$ the voter A is pivotal, six cases for each single voter. In the 24 orderings of type $ABCDX$ the voter D is pivotal, again six cases for each single voter. The bloc is pivotal in the other 72 cases. So the power index for the bloc is $\frac{72}{120} = \frac{3}{5}$, that is $\frac{1}{5}$ for each member, and $\frac{12}{120} = \frac{1}{10}$ for each other voter. Bloc members have twice as much power as the other committee members.

A four-member bloc is pivotal in every case, so each member has power index $\frac{1}{4}$, and the other voters have power index 0.

Practice Exercise. What it the power index for a bloc of size four in a seven-man committee, if five votes are needed to pass a motion?

The number of permutations to be considered becomes much larger as the committee size grows, so the following formulas are useful.

Theorem 4. *Suppose a committee with n members contains a voting bloc of size x, and q votes are required to pass a motion, where $q > \frac{1}{2}n$. Then the Shapley-Shubik power index of the bloc is*

$$\frac{x}{n-x+1} \quad \text{if } x \le n-q+1,$$

$$\frac{n-q+1}{n-x+1} \quad \text{if } n-q+1 \le x \le q,$$

$$\frac{x}{n-x+1} \quad \text{if } q \le x.$$

(In other words, individual members have power index $\frac{1}{n-x+1}$ when $x \le n-q+1$, $\frac{n-q+1}{x(n-x+1)}$ when $n-q+1 \le x \le q$, and $\frac{1}{n-x+1}$ when the bloc size is at least equal to the quota and the bloc has complete control.)

Proof. Suppose we wrote down all the voters in a list from left to right of n members, with the one most favorable to the motion on the left (position 1) and the one least favorable on the right (position n). The pivotal voter will be the one in position q. Say the bloc occupies positions $i, i+1, \ldots, i+x-1$. There are at most $n-x+1$ possible values for i (starting points for the bloc).

When $x \le n-q+1$, the only cases when the bloc will be pivotal is when $i = q-x+1, q-x+2, \ldots, q$, a total of x of the $n-x+1$ possible starting positions. So the fraction in which the bloc is pivotal is $\frac{x}{n-x+1}$.

Suppose $n-q+1 \le x \le q$. The smallest i that will place a bloc member in the pivotal position is again $q-x+1$, but the largest possible value is $n-x+1$, and the number of cases is $(n-x+1) - (q-x+1) + 1 = n-q+1$. So the bloc is pivotal in $\frac{n-q+1}{n-x+1}$ of cases.

If $q \leq x$ then $i + x - 1 \geq q$ is always true, so all positions are pivotal and the index is 1.   □

8.5 General Weighted Systems

Many systems weight some members' votes more heavily than others. Probably the most familiar is a shareholders' vote where each voter receives one vote for each share owned; we brought this up in Chap. 7, in our discussion of cumulative voting.. While some political systems allocate equal numbers of votes to each component, others give more votes to larger states or nations. For example,in the Council of Ministers of the European Union, each nation has one representative, but the number of votes depends on the size of the country: Germany, the United Kingdom and France have 29 votes each, Romania has 14, while Ireland has 7 and Malta has only 3. In order for a proposition to pass, it must receive 74 % of the votes, and at least 50 % of the countries must vote in favor. Many County Boards of Electors in the United States, particularly in New York State, have one member from each city but those from larger cities have more votes.

In a weighted voting system for n participants, suppose the numbers of votes available to the participants (or *weights*) are $w_1, w_2, \ldots, w_n$ and the number of votes required to pass a motion (the *quota*) is q. We shall say the system is of type

$$[q : w_1, w_2, \ldots, w_n].$$

For example, the system used by the committee we discussed in the preceding section was $[3 : 2, 1, 1, 1]$.

Suppose a voter has more votes than the quota. Then that person will be a dictator, as defined previously. In terms of the Shapley-Shubik index, the dictator has power 1 and the others have power 0. On the other hand, consider the type $[6|3, 3, 1]$. The motion will pass if and only if two or three of the first three voters are in favor. The fourth voter is referred to as a *dummy voter*.

Sample Problem 8.2 *What are the Shapley-Shubik power indices of the voters in a system of type* $[11|9, 9, 2]$?

Solution. In any permutation, it is necessary for the first two voters to vote in favor. So each voter has power index $\frac{1}{3}$, even though their vote numbers are very different.

Practice Exercise. What are the Shapley-Shubik power indices of the voters in a system of type $[6|3, 3, 3, 2]$?

In the above example, very different numbers of votes give the same amount of power. But often a change in the number of votes causes a larger change in the power. Here is an example of this.

Sample Problem 8.3 *A voting system is of type* $[5|2, 2, 1, 1]$. *Find the Shapley-Shubik power indices for the four voters.*

Solution. We shall denote the four voters as A, B, X, Y where A and B have two votes each and X and Y have one. The possible voter permutations are shown in the following table, with the pivotal voter highlighted in bold type in each:

AB**X**Y	AX**B**Y	AXY**B**	X**A**BY	XA**Y**B	XY**A**B
BA**X**Y	BX**A**Y	BXY**A**	X**B**AY	XB**Y**A	XY**B**A
AB**Y**X	AY**B**X	AYX**B**	Y**A**BX	YA**X**B	YX**A**B
BA**Y**X	BY**A**X	BYX**A**	Y**B**AX	YB**X**A	YX**B**A

Both A and B have power index $\frac{5}{12}$, while both X and Y have power index $\frac{1}{12}$.

If we make what looks like a small change in the above Worked Example 8.3, allotting one further vote to voter A, the change is again significant. Once more the pivotal voter is highlighted in bold type.

AB**X**Y	AX**B**Y	AXY**B**	X**A**BY	XA**Y**B	XY**A**B
BA**X**Y	BX**A**Y	BXY**A**	X**B**AY	XB**Y**A	XY**B**A
AB**Y**X	AY**B**X	AYX**B**	Y**A**BX	YA**X**B	YX**A**B
BA**Y**X	BY**A**X	BYX**A**	Y**B**AX	YB**X**A	YX**B**A

Now A has power index $\frac{7}{12}$, B has power index $\frac{3}{12}$, while both X and Y again have power index $\frac{1}{12}$.

8.6 Coalitions: The Banzhaf Power Index

In many cases, one expects all members of a certain group to vote the same way, but this is not guaranteed. For example, in United States political houses, we usually expect all Republicans to vote the in favor of issues that follow their party's policies, and against those that oppose them, but sometimes a small number of members will vote otherwise. A group that all plan to support, or all plan to oppose, a motion is called a *coalition*. A *winning coalition* is one that favors the motion in question, and has enough votes that the motion will pass provided all members of the coalition vote for it. The phrase *blocking coalition* is used for one that opposes the motion, and has enough votes to defeat it.

Sometimes there will be one voter with the property that the group will be a winning coalition if that person is a member of it, but not otherwise. We call this a *critical voter*. In other words, a critical voter is a voter who, if he changed his or her vote from yes to no, would cause the measure to fail. For example, consider the earlier example of four voters, one of whom (the Chair, X) can exercise a second vote when necessary, while the other members (A, B, C) have one vote each. If a coalition XA is formed in favor of a motion, then it is a winning coalition, and both X and A are critical voters. If the coalition were XAB or $XABC$ then only X is critical. A coalition with no critical voter is called *invulnerable*. On the other hand, a winning coalition in which every voter is critical is called a *minimal winning coalition*.

In 1946, Lionel Penrose [22] invented a measure of critical voting; it was publicized by John Banzhaf [5] and is usually called the *Banzhaf power index*. To calculate the power of a voter using the Banzhaf index, list all the winning coalitions, then count the critical voters. A voter's Banzhaf power is measured as the fraction of all swing votes that he could cast.

In the example of the four-voter committee, the winning coalitions must all contain at least three votes. Those with three votes are ABC, XA, XB and XC, and in each case all members are critical voters—three for X and two for each of the others. The four-vote coalitions are XAB, XBC and XCA, and in each case X is the only critical voter—three more for X. The final winning coalition, $XABC$, has no critical voters. So the contribution from winning coalitions to the Banzhaf index for X is 6, and to the index for A, B and C is 2 in each case. The data can be represented in the following table:

	ABC	XA	XB	XC	XAB	XBC	XCA	$XABC$	sum
X	0	1	1	1	1	1	1	0	6
A	1	1	0	0	0	0	0	0	2
B	1	0	1	0	0	0	0	0	2
C	1	0	0	1	0	0	0	0	2

The total number of critical voters is 12, so X has power $\frac{6}{12} = \frac{1}{2}$ while A, B and C each have power $\frac{1}{6}$.

We often refer to number of coalitions in which a voter is critical as *points*; in the example, we might say that X received three points from the three-vote winning coalitions and three points from the four-vote ones.

Suppose a proposition requires a quota of q votes, and a certain winning coalition has x members. Necessarily $x \geq q$, and the coalition has $x - q$ more votes than it needs. The critical voters will be the members with more than $x - q$ votes. For the same proposition, a blocking coalition will need at least $(n - q + 1)$ votes, where n

is the total number of votes available, and in a blocking coalition with y members the excess is $y - (n - q + 1)$, so critical voters are those with more than that number of votes.

Sample Problem 8.4 *Consider a five-voter committee in which the Chair, X, has two votes and the other (junior) voters A, B, C and D have one vote each. The quota for passing a motion is four votes. List all winning coalitions. Repeat this for the blocking coalitions. Calculate the Banzhaf power index for each member.*

Solution. A winning coalition must all contain at least four votes. Those with four votes are $ABCD$ and the six coalitions containing X and two others, such as XAB, and in each case all members are critical voters—seven for X and three for each of the others. The four-vote coalitions include $XABC$, $XABD$, $XACD$ and $XBCD$, and in each case X is the only critical voter—four more for X. In the other case, $ABCD$, all voters are critical. The final winning coalition, $XABCD$, is invulnerable. So the contribution to the Banzhaf index for X is 10, and the contribution for A, B, C and D is 4 in each case.

A blocking coalition must contain three or more votes. Those with three votes include four with X and one other voter and four with three of the four junior voters. In each case all voters are critical, a total of four occurrences for each voter. With four votes there are six coalitions with X and two juniors, and the coalition $ABCD$; X is critical in the first six cases, and each junior is critical in the last case. So the contributions to the Banzhaf index from blocking coalitions is again 10 for X, and 4 for each junior.

The Banzhaf power index is $\frac{20}{52} = \frac{5}{13}$ for X and $\frac{2}{13}$ for each of the juniors.

Practice Exercise. Repeat this problem for a three-member committee with a Chair who has two votes, when the quota is three votes.

In the above example, each voter received the same number of points from winning coalitions as from blocking ones. This is not a coincidence.

Theorem 5. *The number of winning coalitions in which a voter is critical is the same as the number of blocking coalitions in which the same voter is critical.*

Proof. Suppose the set S of voters is a winning coalition for a certain proposition and X is a critical voter, and T is the set of all voters not in S. If X changes position and votes against the motion, it will lose, so $T \cup \{X\}$ is a blocking coalition. Conversely, if S is a blocking coalition and X is a critical voter, then if X changes position a winning coalition is formed. So there is a one-to-one correspondence between the winning coalitions and blocking coalitions for which X is critical.   □

Exercises 8

1. A state Senate has 24 members. Motions must be passes by a simple majority. In the event of a 12–12 draw, the Senate President can cast another vote.

   (i) What would constitute a winning coalition or a blocking coalition?

   (ii) What is the Shapley-Shubik power index of the Senate President?

2. In 1958, the Council of Ministers in the European Economic Community (as it was then called) had six members, France, Germany, Italy, the Netherlands, Belgium and Luxembourg, with vote numbers 4, 4, 4, 2, 2, 1 respectively. (See, for example, [9].) The quota was 12. What were the Shapley-Shubik power indices of the countries?

3. In the 1958 Council of Ministers defined in question 2, what are the Banzhaf power indices of the countries?

4. What are the Shapley-Shubik power indices in the example of Sample Problem 8.3, if the quota is changed to 4?

5. What are the Shapley-Shubik power indices for a system of type $[5 : 3, 2, 2, 1]$?

6. What are the Shapley-Shubik power indices for a system of type $[18 : 10, 9, 7, 6]$?

7. What is the power of a voting bloc with 12 members in a committee of size 21 if the quota is

   (i) 12;   (ii) 14;   (iii) 16?

8. Consider a system of type $[19 : 10, 10, 10, 4]$. Show that the voter with weight 4 has no power.

9. A country has the same rules (half, two-thirds) for presidential approval and vetoes as the United States, but its two Chambers have only six members each. What is the Shapley-Shubik power index of that country's President? What are the indices of the members of the two chambers?

10. A weighted voting system has type $[51 : 35, 30, 20, 15]$.

    (i) Calculate the Banzhaf power indices of the four voters.

    (ii) What are the Banzhaf power indices if the quota is changed to 56?

11. A committee has a chairman with three votes and five members with one vote each. The quota for its decisions in five votes. Calculate the Banzhaf power index for each member.

12. A voter has *veto power* when she or he has the power to block any motion. In other words, such a voter forms a one-member blocking coalition. In a weighted voting system of type $[20|8, 8, 6, 4]$, which if any of the voters has veto power?

Solutions to Practice Exercises

Chapter 2

2.1 A receives 12 votes, B receives 11, C receives four. So (i) there is no majority winner (as there are 27 voters, 14 votes would be needed), and (ii) A is the plurality winner. In a primary election, A and B are selected to contest the runoff. For the runoff, C is deleted, so the preference profile is

7	5	8	3	4
A	A	B	B	B
B	B	A	A	A

,

or (combining columns with the same preference list)

12	15
A	B
B	A

.

So B wins the runoff.

2.2 (i) The votes for A, B, C and D are 13, 12, 7 and 11 respectively, so A would win under plurality voting.

(ii) Under the runoff method A and B are retained, and the new preference profile is:

6	7	7	7	2	7	5	2
A	B	A	A	A	B	B	B
B	A	B	B	B	A	A	A

,

giving 22 votes to A and 21 votes to B, so A wins again.

© Springer International Publishing Switzerland 2014
W.D. Wallis, *The Mathematics of Elections and Voting*,
DOI 10.1007/978-3-319-09810-4

(iii) In the Hare method we first eliminate C, obtaining

6	7	7	7	2	7	5	2
A	B	A	D	D	D	B	D
D	D	D	A	A	B	D	B
B	A	B	B	B	A	A	A

Now we eliminate B:

6	7	7	7	2	7	5	2
A	D	A	D	D	D	D	D
D	A	D	A	A	A	A	A

So D wins 30–13.

2.4   Using a $3, 2, 1$ count, the scores are $A : 30, B : 31, C : 17$, and B wins. Under a $4, 2, 1$ count, the totals are $A : 37, B : 36, C : 18$, and A wins.

Chapter 3

3.1   Under the Hare method, C is eliminated, and A beats B 9–4. Also B beats C 10–3, but C beats A 7–6, so there is no Condorcet winner. For Condorcet's solution, we see

$$B \to C(10 - 3), A \to B(9 - 4), C \to A(7 - 6).$$

The first two yield the list ABC and the last result is ignored, so A is elected.

Chapter 4

4.2   Originally, S is eliminated first, giving

8	6	5
P	Q	R
Q	R	P
R	P	Q

Next is R, leaving

8	6	5
P	Q	P
Q	P	Q

and P wins.

If one voter switches, the profile becomes

8	5	1	5
P	Q	R	R
Q	R	Q	S
R	P	P	P
S	S	S	Q

Again S is eliminated, leaving

8	5	1	5
P	Q	R	R
Q	R	Q	P
R	P	P	Q

But now Q is eliminated, giving

8	5	1	5
P	R	R	R
R	P	P	P

and R wins.

In a head-to-head competition between Q and R, Q would win 14–6 in the original election, and 13–7 using the modified preferences.

Chapter 5

5.1 Suppose there are n candidates, and you wish to achieve the overall ranking $A_1, A_2, \ldots, A_n$. Consider the profile in which every voter ranks the candidates $A_1, A_2, \ldots, A_n$. If $i < j$, every voter ranks A_i higher than A_j, so by Pareto efficiency the electorate ranks $A - i$ higher than A_j. So every ranking $A_1, A_2, \ldots, A_n$ can be achieved.

Chapter 6

6.1 There are 32 voters, so the quota is 9, and the surplus is the number of votes by which a candidate exceeds 8. A receives 10 first place votes, so A is elected.

A has a surplus of 2. The votes are divided in proportion 5:5 between A's two lists, so we give one vote to each. After this allocation, the new table is

1	1	7	8	5	2
B	E	B	C	D	E
E	B	D	D	C	B
D	D	E	E	E	C
C	C	C	B	B	D

B has eight votes, C has eight, D has five, and E has three. No one meets the quota. So E (who had the fewest votes) is eliminated. We now have

1	1	7	8	5	2
B	B	B	C	D	B
D	D	D	D	C	C
C	C	C	B	B	D

B has 11 votes and is elected. The surplus of three votes is distributed in ratio $1 : 1 : 7 : 2$, approximately $0.27 : 0.27 : 1.91 : 0.55$. The table is

0.27	0.27	1.91	8	5	0.55
D	D	D	C	D	C
C	C	C	D	C	D

C now has 8.55 votes and is elected. In total, A, B and C are elected.

6.2   As in the previous problem, A has exceeded the quota and is elected; the new profile is

1	1	7	8	5	2
B	E	B	C	D	E
E	B	D	D	C	B
D	D	E	E	E	C
C	C	C	B	B	D

No candidate has received the quota. B, with the most last-place votes, is eliminated; after B's first-place votes are reallocated, the profile is

1	1	7	8	5	2
E	E	D	C	D	E
D	D	E	D	C	C
C	C	C	E	E	D

D is the only candidate to beat the quota, and is elected; the remaining four votes are distributed, approximately, as

1	1	2.3	8	1.7	2
E	E	E	C	C	E
C	C	C	D	D	C

and C wins by 9.7–6.3. So A, C and D are elected.

6.3   Again, A is elected, and there are 12 surplus votes. They are distributed as

3	3	3	3	3	1
B	C	B	C	B	C
C	B			C	B

and B has 9 votes to C's 7. The quota has been achieved, so B is elected. The result is the same under the dynamic system.

Chapter 8

8.1 There are 24 orderings. In six of them the bloc is first, and the next voter, the first of the single voters, is pivotal. So each single voter is pivotal in two cases. In the other 18, the bloc is pivotal. So the individual voters have power $\frac{1}{12}$, and the bloc has power index $\frac{3}{4}$, or $\frac{3}{16}$ for each member. For ease of comparison, this is approximately 18.8 % for bloc members and 8.3 % for others.

8.2 The motion will pass if and only if two of the first three voters are in favor. The fourth voter is a dummy. The other three have equal power, so the indices are $\frac{1}{3}, \frac{1}{3}, \frac{1}{3}, 0$.

8.4 Write X for the chair and A, B for the others. The winning coalitions are X, A and X, B, in which both voters are critical, and X, A, B, in which only X is critical. So X receives 3 points, A and B one each. The blocking coalitions are X and A, B, where each member is critical, so they each receive one point; X, A and X, B, where only X is critical, and X, A, B, which is invulnerable. So X receives three points from the blocking coalitions and the others receive one each. So the points are 6 for X, 2 for A and 2 for B. The indices are $\frac{3}{5}, \frac{1}{5}$ and $\frac{1}{5}$ respectively.

References

1. Arrow, K.J.: A difficulty in the concept of social welfare, J. Political Econ. **58**, 328–346 (1950)
2. Arrow, K.J.: Social Choice and Individual Values, 1st edn. Wiley, New York (1951)
3. Arrow, K.J.: Social Choice and Individual Values, 2nd edn. Wiley, New York (1963)
4. Black, D.: The Theory of Committees and Elections. Cambridge University Press, Cambridge (1958).
5. Banzhaf, J.F.: Weighted voting doesn't work: a mathematical analysis. Rutgers Law Rev. **19**, 317–343 (1965)
6. Barberà, S., Peleg, B.: Strategy-proof voting schemes with continuous preferences. Soc. Choice Welf. **7**, 31–38 (1990)
7. Benoît, J.-P.: The Gibbard–Satterthwaite theorem: a simple proof. Econ. Lett. **69**, 319–322 (2000)
8. Bonner, A.: Doctor Illuminatus: A Ramon Llull Reader. Princeton University Press, Princeton (1993)
9. Brams, S.J.: Paradoxes in Politics: An Introduction to the Non-obvious in Political Science. Free Press, New York (1976)
10. Brams, S.J., Fishburn, P.C.: Approval Voting, 2nd edn. Springer, New York (2007)
11. de Condorcet, M.: Essai sur l'Application de l'Analyse à la probabilité des décisions rendues à la pluralité des voix. Imprimerie Royale, Paris (1785)
12. Coombs, C.H.: A Theory of Data. Wiley, New York (1964)
13. Droop, H.R.: On methods of electing representatives. J. Stat. Soc. Lond. **44**, 141–196 (1881). (With discussion, 197–202), Reprinted in Voting Matters **24**, 7–46 (2007)
14. Fishkin, J.S.: The Voice of the People: Public Opinion and Democracy. Yale University Press, New Haven (1964)
15. Geanakoplos, J.: Three brief proofs of Arrow's impossibility theorem. Econ. Theory **26**, 211–215 (2005)
16. Gibbard, A.: Manipulation of voting schemes: a general result. Econometrica **41**, 587–600 (1973)
17. http://history.house.gov/Institution/Presidential-Vetoes/Presidential-Vetoes/
18. May, K.: A set of independent necessary and sufficient conditions for simple majority decisions. Econometrica **20**, 680–684 (1952)
19. McLean, I., Nanson, E.J.: social choice and electoral reform. Aust. J. Pol. Sci. **31**, 369–385 (1996)
20. Ninjbat, U.: Another direct proof for the Gibbard–Satterthwaite theorem. Econ. Lett. **116**, 418–421 (2012)
21. http://www.ohrt.com/odds/binomial.php
22. Penrose, L.: The elementary statistics of majority voting. J. R. Stat. Soc. **109**, 53–57 (1946)

© Springer International Publishing Switzerland 2014
W.D. Wallis, *The Mathematics of Elections and Voting*,
DOI 10.1007/978-3-319-09810-4

23. http://rangevote.com/
24. http://www.rapidtables.com/calc/math/Factorial_Calculator.htm
25. Satterthwaite, M.A.: Strategy-proofness and Arrow's conditions: existence and correspondence theorems for voting procedures and social welfare functions. J. Econ. Theory **10**, 187–217 (1975)
26. Shapley, L.S., Shubik, R.: A method for evaluating the distribution of power in a committee system. Am. Pol. Sci. Rev. **48**, 787–792 (1954)
27. Solgård, T.A., Landskroener, P.: Municipal voting system reform: overcoming the legal obstacles. http://www.mnbar.org/benchandbar/2002/oct02/voting.htm
28. Straffin, P.D., Jr.: Topics in the Theory of Voting. Birkhäuser, Boston (1980)
29. Taylor, A.: The manipulability of voting systems. Am. Math. Mon. **109**, 321–337 (2002)
30. Taylor, A.: A paradoxical Pareto frontier in the cake-cutting context. Math. Soc. Sci. **50**, 227–233 (2005)
31. Taylor A.D., Pacelli, A.M.: Mathematics and Politics, 2nd edn. Springer, (2008)
32. Wikipedia. http://en.wikipedia.org/wiki/Comparison_of_the_Hare_and_Droop_quotas
33. Wikipedia. http://en.wikipedia.org/wiki/Cumulative_voting
34. Wikipedia. http://en.wikipedia.org/wiki/Electoral_College_(United_States)
35. Wikipedia, http://en.wikipedia.org/wiki/Range_voting
36. Woodall, D.R.: Monotonicity and single-seat election rules. Voting Matters **6**, 9–14 (1996)